LEARNING
LANDSCAPE
ECOLOGY

LEARNING LANDSCAPE ECOLOGY

A PRACTICAL GUIDE TO CONCEPTS AND TECHNIQUES

SARAH E. GERGEL
University of British Columbia

MONICA G. TURNER
University of Wisconsin, Madison

Editors

 Springer

Sarah E. Gergel
Department of Forest Sciences
University of British Columbia
Vancouver, British Columbia
Canada V6T 1Z4
sarah.gergel@ubc.ca

Monica G. Turner
Department of Zoology
University of Wisconsin
Madison, WI 53706
USA
mgt@mhub.zoology.wisc.edu

Library of Congress Cataloging-in-Publication Data
Learning landscape ecology : a practical guide to concepts and techniques / edited by Sarah E. Gergel, Monica G. Turner.
     p.   cm.
   Includes bibliographical references and index.
   ISBN 0-387-95254-3 (sc)
   1. Landscape ecology.   I. Gergel, Sarah E.   II. Turner, Monica Goigel.
   QL541.15.L35 L43   2002
   577—dc21   00-069118

Printed on acid-free paper.

ArcExplorer™ and the GIS by ESRI emblem are trademarks provided under license from Environmental Systems Research Institute, Inc.

PREFACE

Landscape ecology continues to grow as an exciting, dynamic ecological discipline. With its broadscale emphasis and multidisciplinary approach, landscape ecology lends itself both to basic research and to applications in land management, land-use planning, wildlife management, ecosystem management, and conservation biology. Landscape ecology makes a unique contribution to the scientific community in its attention to ecological dynamics across a broad range of spatial and temporal scales, and as a result it has become increasingly important for students in the natural sciences to gain a basic understanding of the subject. Colleges and universities across the United States are incorporating courses in landscape ecology into their curricula. However, nearly every book on landscape ecology is a book to be *read*, lacking a hands-on approach. This text is intended to fill that void by providing a comprehensive collection of landscape ecology laboratory exercises.

These teaching exercises stress the fundamental concepts of landscape ecology, rather than highly specialized, technical methods. While students will gain experience using a variety of tools commonly used in landscape ecology, we stress the conceptual understanding necessary to use these techniques appropriately. This book attempts to convey the myriad approaches used by landscape ecologists (as well as a multitude of approaches to teaching) and include group discussion, thought problems, fieldwork, data analysis, spatial data collection, exposure to Geographic Information Systems (GIS), simulation modeling, analysis of landscape metrics, spatial statistics, and written exercises.

This book is divided into seven sections, which complement the companion textbook, *Landscape Ecology in Theory and Practice: Pattern and Process*, by M. G. Turner, R. H. Gardner, and R. V. O'Neill. However, this textbook is also a useful stand-alone volume that can be used for teaching and learning. We also hope practicing landscape ecologists will find this book to be a useful reference.

This book provides labs spanning a range of difficulty levels; thus, we have provided "Suggestions for Instructors," which include difficulty ratings as well as several suggested lab sequences for undergraduate- and graduate-level courses. Many of the exercises also require computers. However, we intentionally designed the labs to be very user friendly in a PC environment. The enclosed CD-ROM includes files for many of the labs, including easily installable computer programs, simple DOS executable files, and data files accessible using commonly available word processing or spreadsheet programs. The CD also includes color images that are viewable using a web browser or Adobe Acrobat Reader software (available free on the web at www.adobe.com). Our intent is to make these teaching materials readily usable by colleges and universities without elaborate computing facilities.

This volume focuses on computer-oriented labs more than field-oriented labs, even though field studies are a critical component of landscape ecology. There are several reasons for this emphasis. First, designing exercises that are transportable to *any* landscape is tricky because the patterns, relevant scales, and important biological processes differ among landscapes (however, two chapters involving fieldwork are included). Field-oriented labs, in general, may be best designed by the instructor to emphasize the important features and dynamics of a particular local area. Second, the quantitative techniques developed during the past decade or so in landscape are unfamiliar to many students, and thus this book fills an important void. Students have few opportunities to use simulation models and to apply the methods of spatial analyses. We hope this volume will help share the quantitative and modeling expertise of a few within an even broader scientific audience. However, the quantitative emphasis in this book is in no way intended to diminish the important role that fieldwork plays in landscape ecology. We strongly encourage students and instructors to embark on field studies in their local landscapes.

Acknowledgments

So many people were fundamental to this endeavor and deserve our praise and thanks. First, we'd like to thank all the contributing authors for their creativity and enthusiasm in this endeavor, as well as their flexibility in allowing their chapters to be molded into part of the "whole." Second, the following external reviewers deserve huge thanks for their critical assessment and enthusiasm: Tim Allen, Matthias Burgi, Jiquan Chen, Jonathan Chipman, Graeme Cumming, Don DeAngelis, Amy Downing, Mike DeMers, Curtis Flather, Marie-Josée Fourtin, Frank Golley, Steven Hamburg, Andy Hansen, Tom Hoctor, Lou Iverson, Tony Ives, Jeffrey Klopatek, David Lewis, Nancy

Matthews, Kevin McGarigal, Nancy McIntyre, Todd Miller, Ron Moen, Kirk Moloney, Barry Noon, Bob O'Neill, Volker Radeloff, Marguerite Remillard, Kurt Riitters, Tania Schoennagel, Fred Sklar, Pat Soranno (and her lab), Tom Spies (and his lab), Dean Urban, Steve Ventura, Karen Whitney, John Wiens, and the lab group of David Mladenoff. Rebecca Reed also deserves special thanks for her contributions during the beginning stages of the book.

Equally important were the wide variety of students who tested these labs, helped find our mistakes, and offered suggestions for improvement. First and foremost, Sarah's graduate students in the fall 1998 landscape ecology lab course at University of Wisconsin, Madison, Jill Bukovac, Bruce Kahn, David Lewis, and Anna Pidgeon, tested 13 of the labs in one semester. Students in several years of Monica Turner and David Mladenoff's landscape ecology classes at University of Wisconsin, Madison, also deserve special thanks for their patience with the first drafts of many of these labs. We were continually impressed with their thorough evaluations and insightful, constructive suggestions. Instructors at other colleges and universities (and their students who are too numerous to mention here!) also deserve many thanks for testing labs, including Tara Reed's undergraduate biology class at Lawrence College and Joshua Greenberg's landscape ecology course at the University of Washington, Seattle, and Cheryl Schultz's graduate course in Landscape Ecology at The University of California, Santa Barbara. Of course, the many students enrolled in the courses taught by chapter authors deserve much thanks for their role in the development of these labs. Sharon Cowling, Lisa Dent, Dan Kashian, and Jack Williams, and all the post-docs at NCEAS who participated in the "Eigenbeer challenge" deserve special thanks for their attention to detail in the proofing stage of the book. Lastly, the artistic prowess of Michael Turner and Dirk Brandts in improving the figures in this book can not be overstated.

Finally, we would like to thank everyone in the Turner lab, especially Matthias Bürgi, Jeff Cardille, Mark Dixon, Dan Kashian, Tania Schoennagel, Mark Smith, and Dan Tinker, for many discussions, critiques, and help with endless aspects of this book. Sarah would also like to thank the recently retired Larry D. Harris for introducing her to the field of landscape ecology as an undergraduate. Sarah also thanks Jon Shurin for feeding her, enabling her to simultaneously finish this book and her Ph.D. We hope that this book will help aspiring landscape ecologists use the concepts and tools of landscape ecology to further our knowledge of how landscapes function and change, and more importantly, build and expand on the ideas presented here.

Sarah E. Gergel
Monica G. Turner
Madison, Wisconsin

SUGGESTIONS FOR INSTRUCTORS

http://www.nceas.ucsb.edu/LearningLandscapeEcology

This book explores a variety of topics in landscape ecology, and as result, the difficulty levels of the labs also vary greatly. We hope this will provide a range of teaching tools appropriate for undergraduates, beginning graduate students, and advanced graduate students and researchers. We have assessed the difficulty of the labs using three categories:

Undergraduate—first introduction to landscape ecology, less emphasis on quantitative or computer-oriented labs, assumes a semester of ecology

Graduate—assumes at least one semester of statistics and a basic familiarity with a Windows PC and common word processing and spreadsheet software

Advanced Graduate—assumes experience in conducting research in landscape ecology, solid facility with PCs, and beginning familiarity with some aspects of modeling

These categories assume a 3-hour class period. Thus, while a graduate lab might be used with undergraduates, extra time for completion should be allowed; when using an undergraduate lab with graduates, it will likely be completed in less than 3 hours. Many labs also include discussion questions and write-up sections to be completed out of class.

Some authors have also created more detailed "Instructor's Notes" for several chapters. Please see our website for availability (as they are subject to revision as questions arise). We also strongly recommend that the instructor gain familiarity with the labs before use in the classroom. These labs have been tested extensively; however, unforeseen computer glitches may arise when using the programs on a different computer, with a different operating system, and so forth. Check our website for corrections or to report problems.

Software Requirements

Chapters 1, 2, 6, 10, 11, 12, 13, 14, 17, 18 and part of 7 use only Excel (.xls) files, Adobe (.pdf) files, or no files at all, and as such are compatible with Macintosh or Windows platforms which have Excel and/or Adobe Acrobat Reader installed. Adobe Acrobat Reader is available for free from the web at:
   http://www.adobe.com
Several other programs require a PC running a recent version of Windows, or require a Windows emulator for use on a Macintosh platform: Markov, HarvestLite, Rule, Fragstats, ReserveDesign, Folio, ArcExplorer,™ and Bachmap. All programs are compatible with Windows XP except for Fragstats 2.0.

Fragstats 2.0 and Windows XP

The version of Fragstats included on the CD (FRAGSTATS 2.0) is not compatible with Windows XP; however, it does work with Windows 95/98/2000. Using Fragstats on Windows XP requires downloading the latest shareware version of Fragstats 3.0 from the web at:
   http://www.umass.edu/landeco/research/fragstats/fragstats.html
Be sure to visit our website before you begin teaching (or if you experience any difficulties while teaching) as it is frequently updated with corrections and helpful tips:
   http://www.nceas.ucsb.edu/LearningLandscapeEcology

Some Suggested Course Sequences

Undergraduate Capstone Course in Landscape Ecology/Land Management/Conservation Biology
   Introduction to Geographic Information Systems (GIS)
   Simulating Changes in Landscape Pattern
   Interpreting Landscape Patterns from Organism-Based Perspectives
   Landscape Context
   Modeling Ecosystem Processes (Basic Version)
   Reserve Design

Graduate Course in Landscape Ecology
   Scale and Hierarchy Theory
   Collecting Spatial Data at Broad Scales
   Creating Landscape Pattern
   Introduction to Markov Models

Understanding Landscape Metrics I

Neutral Landscape Models

Landscape Disturbance: Location, Pattern, and Dynamics

Individual-Based Modeling: The Bachman's Sparrow

Modeling Ecosystem Processes

Feedbacks between Organisms and Ecosystem Processes

Prioritizing Reserves for Acquisition

Advanced Graduate Course or Spatial Modeling

Introduction to Markov Models

Simulating Changes in Landscape Pattern

Understanding Landscape Metrics II: Effects of Changes in Scale

Scale Detection Using Semivariograms and Autocorrelograms

Alternative Stable States

Landscape Connectivity and Metapopulation Dynamics

Modeling Ecosystem Processes (Advanced Version)

Chapter	Title	Author	Difficulty Level	Related Chapters in *Landscape Ecology in Theory and Practice: Pattern and Process*, by M. G. Turner, R. H. Gardner, and R. V. O'Neill
1	Scale and Hierarchy Theory	O'Neill	Undergraduate; short discussion lab	2
2	Collecting Spatial Data at Broad Scales	Gergel et al.	Graduate; requires significant out-of-class work	1, 5
3	Introduction to Geographic Information Systems (GIS)	Greenberg et al.	Undergraduate	
4	Introduction to Markov Models	Urban and Wallin	Graduate	3
5	Simulating Changes in Landscape Pattern	Gustafson	Undergraduate or Graduate	3, 4
6	Creating Landscape Pattern	Delcourt	Includes different versions for Undergraduate or Graduate; Understanding Landscape Metrics I is pre-requisite for Advanced version	4
7	Understanding Landscape Metrics I	Cardille and Turner	Graduate	5
8	Understanding Landscape Metrics II: Effects of Changes in Scale	Greenberg et al.	Graduate; essential prerequisites: Understanding Landscape Metrics I, Introduction to Geographic Information Systems	5

(*Continued*)

CONTENTS

CONTRIBUTORS

Jeffrey A. Cardille
Department of Zoology
University of Wisconsin, Madison
Madison, WI 53706
cardille@facstaff.wisc.edu

John R. Cary
Department of Wildlife
   Ecology
University of Wisconsin, Madison
Madison, WI 53706
jrcary@facstaff.wisc.edu

F. S. (Terry) Chapin, III
Institute of Arctic Biology
University of Alaska, Fairbanks
Fairbanks, AK 99775
fschapin@lter.uaf.edu

Hazel R. Delcourt
Department of Ecology and
   Evolutionary Biology
University of Tennessee
Knoxville, TN 37996
hdelcourt@utk.edu

John B. Dunning, Jr.
Department of Forestry and
   Natural Resources
Purdue University
West Lafayette, IN 47907
bdunning@fnr.purdue.edu

Jerry F. Franklin
College of Forest Resources
University of Washington
Seattle, WA 98195
jff@u.washington.edu

Robert H. Gardner
Center for Environmental Science
Appalachian Laboratory
Frostburg, MD 21532
gardner@al.umces.edu

Sarah E. Gergel
Department of Forest Sciences
University of British Columbia
Vancouver, British Columbia
Canada V6T 1Z4
sarah.gergel@ubc.ca

Stephen T. Gray
Department of Botany
University of Wyoming
Laramie, WY 82701
sgray@uwyo.edu

Joshua D. Greenberg
Skagit County GIS/Mapping
Mount Vernon, WA 98273
joshg@co.skagit.wa.us

Eric J. Gustafson
North Central Research Station
USDA Forest Service
Rhinelander, WI 54501
egustafson@fs.fed.us

Jianguo (Jack) Liu
Department of Fisheries and Wildlife
Michigan State University
East Lansing, MI 48824
jliu@perm3.fw.msu.edu

Miles G. Logsdon
School of Oceanography
University of Washington
Seattle, WA 98195
mlog@u.washington.edu

David J. Mladenoff
Department of Forest Ecology and
   Management
University of Wisconsin, Madison
Madison, WI 53706
djmladen@facstaff.wisc.edu

Robert V. O'Neill, retired
Environmental Sciences Division
Oak Ridge National Laboratory
Oak Ridge, TN 37831
eoneill@attglobal.net

Michael W. Palmer
Department of Botany
Oklahoma State University
Stillwater, OK 74078
carex@osuunx.ucc.okstate.edu

Scott M. Pearson
Biology Department
Mars Hill College
Mars Hill, NC 28754
spearson@mhc.edu

Garry D. Peterson
Department of Geography &
   McGill School of the Environment
McGill University
805 Sherbrooke St. W.
Montreal, Quebec
Canada H3A 2K6

Tara Reed-Andersen
Department of Natural and
   Applied Sciences
University of Wisconsin, Green
   Bay
Green Bay, WI 54311
reedandt@uwgb.edu

Mark A. Smith
Department of Wildlife Ecology
   and Zoology
University of Wisconsin, Madison
Madison, WI 53706
masmith7@students.wisc.edu

David J. Stewart
Department of Anthropology
University of Georgia
Athens, GA 30602
dstewart@julian.dac.uga.edu

Stanley A. Temple
Department of Wildlife
   Ecology
University of Wisconsin, Madison
Madison, WI 53706
satemple@facstaff.wisc.edu

Daniel B. Tinker
Department of Botany
University of Wyoming
Laramie, WY 82071
Tinker@uwyo.edu

Monica G. Turner
Department of Zoology
University of Wisconsin, Madison
Madison, WI 53706
mgt@mhub.zoology.wisc.edu

Dean L. Urban
Nicholas School of the
   Environment
Duke University
Durham, NC 27708
deanu@pinus.env.duke.edu

Linda L. Wallace
Department of Botany and
   Microbiology
University of Oklahoma
Norman, OK 73019
lwallace@ou.edu

David O. Wallin
Center for Environmental Science
Huxley College of Environmental
   Studies
Western Washington University
Bellingham, WA 98225
wallin@cc.wwu.edu

Steven Walters
U.S. Environmental Protection
   Agency
Narragansett, RI 02882
Walters.Steve@epamail.epa.gov

Kimberly A. With
Division of Biology
Kansas State University
Manhattan, KS 66506
kwith@ksu.edu

INTRODUCTION AND CONCEPTS OF SCALE

The concept of scale is fundamental to ecology, and the importance of scale has been emphasized strongly in landscape ecology. Furthermore, scale-related terminology can be complicated. Chapter 1, Scale and Hierarchy Theory, lays a fundamental framework for addressing issues of scale by defining the basic components of scale and distinguishing between scale and levels of organization. One reason scale is so important to landscape ecologists is that spatial data are often derived from different data sources and mapped at different scales. Chapter 2, Collecting Spatial Data at Broad Scales, introduces several different spatial data sources used by landscape ecologists, including aerial photographs, topographic maps, satellite imagery, and field-collected data. In Chapter 2, students compare and contrast the results obtained from these various data sources and examine the trade-offs inherent to different data types. A widely used tool for viewing and analyzing spatial data is a Geographic Information System (GIS), which has helped shape the way landscape ecologists ask and answer questions. Chapter 3, Introduction to GIS, presents an extremely user-friendly guide to understanding GIS and provides experience using ArcExplorer, a useful starting point for anyone interested in GIS technology.

CHAPTER 1

SCALE AND HIERARCHY THEORY

ROBERT V. O'NEILL AND MARK A. SMITH

OBJECTIVES

Many aspects of ecological processes change with the scale at which they are observed. As a result, understanding how scale influences our observations is critical to understanding ecology, particularly landscape ecology. Hierarchy theory provides a framework for examining scale-dependent processes and their resulting patterns. The primary goals of this lab are to

1. illustrate the conceptual differences between the two primary aspects of scale: grain and extent;
2. examine the effects of changes in grain and extent for data collection and interpretation of results;
3. illustrate and explain the implications of hierarchy theory for landscape ecology; and
4. foster an understanding of hierarchical controls on ecological processes (that context is derived from a higher level of organization, whereas mechanistic explanations originate from a level below).

In this lab you will investigate the fundamental concepts of scale and hierarchy theory through some thought exercises and pen-and-paper exercises that work well in a discussion format.

INTRODUCTION

Consider a table such as the one you are likely seated at. Is it "solid"? You cannot put your finger through it. However, if we observed the table at the ultrafine scale of molecules, the table is almost completely empty space—it is not "solid" at all. Changing the scale of observation can change a fundamental property of an object, such as whether it appears solid. Does the table remain the same thickness? If you measured the table every five minutes for the next hour, you would conclude that it doesn't change. If you measured it every century for the next thousand years, you would discover that the friction of elbows and lab manuals is causing the table to become thinner. Thus, changing the scale of observation can change our impression of the fundamental dynamics of an observation set (i.e., the table).

What Is Scale?

Scale is measured by two factors: grain and extent. The **grain** is determined by the finest level of resolution, or measurement, made in an observation. When observing a landscape, the spatial grain might be set by the finest resolution of the remote imagery of the landscape (e.g., a 30 × 30-m pixel). The spatial **extent** of an observation set is established by the total area sampled. As you saw in the table example, significantly increasing or decreasing the grain and extent used to make an observation may influence how an entity appears, or the conclusions one draws about the dynamics of an observation set.

   Hierarchy theory provides a context for examining relationships that change with scale. One of the most important lessons from hierarchy theory is how phenomena change when you alter the scale at which you are observing the ecological system. Next, we examine the effects of changes in scale.

EXERCISE 1
Implications of Changes in Scale

Consider a predator/prey example (inspired from a fish example by Rose and Leggett, 1990) with two species of insects found in leaf litter in forests. Both the predator and prey are sampled in $0.1m^2$ areas of litter throughout a forest stand (thus, the grain is $0.1m^2$). Measurements are taken every 10 meters for a total of 100 meters (establishing the spatial extent of the observation set). The number of individuals of both predator and prey found in each sample are reported in Table 1.1.

   Prepare a graph of the relationship between the predator and the prey. What do you observe? Can you offer an explanation?

   At this fine scale we observe "predator avoidance." If a predator is in the neighborhood, the prey moves away.

   Now change the scale of the sampling. We'll change the extent of the

TABLE 1.1
ABUNDANCE OF INSECTS SAMPLED IN 0.1M² QUADRATS OF FOREST
LEAF LITTER AT 10-METER INTERVALS OVER 100 METERS

Meters	Number of Predators	Number of Prey
0	40	2
10	30	3
20	40	0
30	25	4
40	10	29
50	10	22
60	15	18
70	20	8
80	15	10
90	35	1
100	25	2

data by sampling every 2000 meters over a total length of 20,000 meters (Table 1.2). Graph the data reported in Table 1.2. What do you observe?

At this broader scale, we are sampling over different habitat types throughout the landscape, including agricultural areas, meadows, and other areas bereft of litter. Because both insects require the cover and detritus of leaf litter, both predator and prey are more prolific in forested areas with leaf litter, and less prolific in habitat types with less litter.

The important thing to note is that changing the scale of observation (the grain or the extent of the data set) changed the dominant phenomena controlling the pattern, from predator avoidance at fine scales to land-cover changes at broad scales. What are some other ecological examples in which a change in the scale of observation might change the dominant phenomena governing a pattern? Be prepared to discuss at least two examples.

Hierarchical Structure and Its Implications

The hierarchical structure of ecological systems has important ramifications for how we explain phenomena (Allen et al., 1987). The answer to *why?* (why does something occur?) will ordinarily be found at the next lower level of organization. The answer to *so what?* (what is the significance?) will ordinarily be found at the next higher level of organization. The next example with a Frisbee game will illustrate this.

Imagine that, on the way to class, you pass a group playing Frisbee. One player keeps dropping the Frisbee and you want to explain this. First you establish an observation set = Klutz (Figure 1.1).

TABLE 1.2
ABUNDANCE OF INSECTS SAMPLED IN 0.1M² QUADRATS PLACED
EVERY 2000 METERS OVER A LENGTH OF 20,000 METERS ACROSS
THE LANDSCAPE

Meters	Number of Predators	Number of Prey
0	20	40
2000	12	20
4000	5	1
6000	15	30
8000	0	1
10,000	1	1
12,000	1	8
14,000	27	40
16,000	10	15
18,000	5	9
20,000	20	22

To explain *why* the Frisbee is dropped, you must look within Klutz, at the next lower level of organization (Figure 1.2).

You discover a defect in the nervous system of Klutz that affects his/her muscular coordination, and your question is answered.

To answer the question *so what?* you must look at the larger system Klutz is part of, the Frisbee team (Figure 1.3). You discover that Klutz's Frisbee team loses the game.

Now, let's translate this into the context of landscape ecology. You and a friend are walking hand in hand through the woods and you come upon an isolated clearing. The second thought that crosses your mind is to seek an explanation for the clearing. How should you proceed? Consider the observation set, the *why*, and the *so what?* Be prepared to discuss your rationale.

Hierarchical Levels of Organization

Hierarchy theory postulates that ecological systems are structured in discrete **levels of organization.** In the Frisbee example, Klutz was a lower level of or-

FIGURE 1.1
A diagrammatic representation of the observer's eye and observation set = Klutz

FIGURE 1.2
The perspective of the observer's eye shows the examination of the next lower level of organization within Klutz

ganization relative to the Frisbee team to which (s)he belonged. A level of organization can be examined at a variety of scales. However, once we examine a particular biological entity (hence, making an observation set), we impose on the set a particular scale of observation. Next, you will brainstorm following a set of examples designed to illustrate the differences between, and independent nature of, scale and level of organization, and how formulating a question forces us to set a scale for our observation set.

Consider a snapping turtle (*Chelydra* spp.) sitting on a log in a lake. One hierarchy in which the turtle resides is its taxonomic hierarchy: kingdom: Animalia, phylum: Chordata, class: Reptilia, order: Testudines, family: Chelydridae, genus: *Chelydra*. Another hierarchy in which the turtle resides is the area in which the turtle lives: from the entire state to the county, park, lake, and log on which it sits.

One might wonder why turtles are found in certain areas of the lake. The answer (*why*) may be found at a lower level of organization—because turtles are poikilotherms, they might be found basking in the sun on a log. One might also ask, what controls the distribution of turtles over a broader area, such as the entire state? One might discover that pollution levels in lakes throughout the state govern which lakes contain turtles, as contaminants may cause abnormal development of turtle eggs.

Certainly the significance of turtle distributions (*so what*) can be found at the next higher level of organization. Considering its taxonomic level, it might be important to know if this species is the only representative of its genus in the area. At the higher level of organization of "the lake," does the location of the turtle have implications for other organisms in the lake (e.g., fish and aquatic invertebrates)? Might they avoid the area in which the turtle resides?

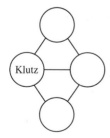

FIGURE 1.3
The observation set Klutz, and Klutz's relationship to a larger context, the frisbee team

EXERCISE 2
Contrasting Hierarchical Levels of Organization and Scale

Next, consider some ecological entity of interest to you. Formulate a question (and hypothesize an answer) regarding its distribution, abundance, behavior, or dynamics. Be prepared to explain how the explanation is related to the next lower hierarchical level, and how the significance is related to the next higher level of organization. Then, ask your question at a different spatial scale, changing either the grain or extent of your observation set. Again, explain the *why* and the *so what* and whether either changed with the scale of your observation.

CONCLUSIONS

In some fields of biology, such as medicine, hierarchical levels are well defined (e.g., cell, organ, body), and problems of scale may seldom arise. If you want to study blood cells, you reach for a microscope. A landscape is less easily defined, however, and scale must be carefully considered in problem formulation, data collection, and analysis of results. Otherwise, you might reach for a magnifying glass when you really need a telescope.

BIBLIOGRAPHY

Note. An asterisk preceding the entry indicates that it is a suggested reading.

*ALLEN, T. F. H. 1998. The landscape "level" is dead: Persuading the family to take it off the respirator. In D. L. Peterson and V. T. Parker, eds. *Ecological Scale: Theory and Applications.* Columbia University Press, New York, chapter 3. Distinguishes levels of organization and scale.

ALLEN, T. F. H., R. V. O'NEILL, AND T. HOEKSTRA. 1987. Interlevel relations in ecological research and management: Some working principles from hierarchy theory. *Journal of Applied Systems* 14 63–79.

*KING, A. W. 1997. Hierarchy theory: A guide to system structure for wildlife biologists. In J. A. Bissonette, ed. *Wildlife and Landscape Ecology: Effects of Pattern and Scale.* Springer-Verlag, New York, pp. 185–212. One of the best concise overviews of hierarchy theory, presented in terms easily accessible to biologists.

*O'NEILL, R.V., D. L. DeANGELIS, J. B. WAIDE, AND T. F. H. ALLEN. 1986. *A Hierarchical Concept of Ecosystems.* Princeton University Press, Princeton, New Jersey. This classic book lays out the conceptual basics of hierarchy theory as it applies to many areas of ecology.

ROSE, G. A., AND W. C. LEGGETT. 1990. The importance of scale to predatory-prey spatial correlations: An example of Atlantic fish. *Ecology* 71:33–43.

*URBAN, D. L., R. V. O'NEILL, AND H. H. SHUGART JR. 1987. Landscape ecology. *BioScience* 37:119–127. A classic early paper that discusses landscape ecology with an emphasis on hierarchical structure.

CHAPTER 2

COLLECTING SPATIAL DATA AT BROAD SCALES

SARAH E. GERGEL, MONICA G. TURNER, AND DAVID J. MLADENOFF

OBJECTIVES

Spatial data are routinely used by landscape ecologists to formulate hypotheses, examine trends in landscape patterns, and make management decisions. Thus, a basic familiarity with the variety of data sources currently available, and an understanding of differences and similarities among them, is a fundamental part of landscape ecology. The goals of this lab are to

1. demonstrate the methods used to obtain spatial data at broad scales;
2. illustrate the differences among and the limitations of different data sources;
3. convey the challenges of collecting and using spatially explicit data; and
4. combine field and laboratory results to illustrate the connections between ground-level data, remote-sensing data, topographic maps, and Geographic Information System (GIS) data.

As a class, students will collect reference data at several field sites selected by the instructor. The data obtained from field sampling will be compared in the laboratory to data collected from other data sources such as aerial photos, topographic maps, satellite imagery, and GIS layers available for your area.

INTRODUCTION

Spatial data commonly used in landscape ecology come from a variety of sources, such as field sampling, aerial photos, topographic maps, satellite images, or an existing GIS. The spatial data from these sources are created using different techniques, have their own set of inherent assumptions, and may accentuate or minimize certain landscape features. Different spatial data types also have different sources of error and provide information at different levels of resolution. Thus, the first step in using spatial landscape data often involves verifying the accuracy of the different sources of data as well as determining which sources fit the needs of the project at hand.

Accuracy assessment involves verifying the accuracy and legitimacy of spatial data against a reliable source of reference data. Field-collected reference data can also be used *a priori* in the preparation of spatial data. Here, we simplify the procedure in order to give you exposure to as many different types of data sources as possible. The class will work in small groups (three to four people) to collect field data at different sites throughout the local area. The field results will then be compared to data collected from selected spatial data sources in the laboratory.

MATERIALS

For fieldwork, you will need the provided **Summary Data Sheet** (Table 2.1 on the CD), additional paper for field notes, a pencil, and a road map. A vehicle will be needed as the transects will span several kilometers. Your instructor will provide a map outlining the study area of each group. For the laboratory work, you will use the **Summary Data Sheet,** a pencil, a ruler, a calculator, and the data sources provided by your instructor.

EXERCISE 1
Fieldwork

DATA COLLECTION

Each group will receive a map outlining the boundaries of their study area of approximately 25 km². Within the study area, you must position three separate transects, each 3 miles in length (depending on the country you are in and your car's odometer, you may wish to sample transects 5 km in length). Ideally, the transects should be fully representative of the cover types in your area. Therefore, the different transects should be established in different cover types within the area, such as suburban, rural, and natural areas.

For data collection, use the odometer to record the length of each cover type on *one side of the road* as you drive along each transect. Your instructor will provide directions regarding which cover type categories to use. In the U.S. Midwest, for example, land-cover categories might include agriculture, urban, forest, wetland, prairie, and water (as shown in the Summary Data Sheet). While it may be necessary to make more detailed sub-categories, they must all aggregate up to the basic cover categories provided by the instructor. A digital version of the **Summary Data Sheet** has been provided on the CD (Table 2.1 under the directory for this lab), which can be altered to suit your particular categories.

It will also be helpful to take very detailed notes of the location of each transect and note any landmarks (e.g., crossroads) or other important identifying features. This information will be crucial in helping to relocate the transects using the other spatial data sources later. In order to ensure that the appropriate spatial data will be available for the areas sampled, *be certain to stay within the boundaries of your assigned area*.

NOTE: As you sample, you will be forced to make decisions and make assumptions about your data, your methodology, your categorizations, and so on. There is not one correct way to sample. Rather, you must consider your purpose for data collection, clearly document your rationale, and *most important, be consistent*!

DATA ANALYSIS

After the data collection from your field transects is complete, each group will be responsible for calculating the following summary statistics *for each transect in their area*:

1. Proportion (*p*) of the total length of each transect occupied by each cover type
2. Mean segment length for each cover type
3. Edges, or the number of times you cross a boundary between two different cover types
4. Coefficient of variation (% CV) for each variable (*p*, mean segment length, and edges) for each cover type, across all data sources

The following formulas may be helpful, where *n* is the number of segments and *x* is any variable (*p*, mean segment length, or edges):

Standard Deviation (s)	Mean ($\bar{x}$)	Coefficient of Variation (%)
$\sqrt{\dfrac{n\sum x^2 - \left(\sum x\right)^2}{n(n-1)}}$	$\dfrac{\sum x}{n}$	$\dfrac{s}{x} \times 100$

Results will be reported using **Table 2.1, Summary Data Sheet,** on the CD. These statistics will be computed again after resampling the same transects using the other data sources.

EXERCISE 2
Spatial Data Sources

Next, in the laboratory, the data collected at your field site will be compared to other commonly used sources of landscape data. Your task is to locate the same transects you sampled in the field and then resample them using the provided spatial data sources. The same summary data will be calculated for each cover type on each transect. Use the following sources of data provided by your instructor (as available).

Topographic Maps

Maps are graphic representations of the earth's surface and are based on a set of assumptions and decisions as to what constitutes "important" information. This is governed, to a large extent, by the scale of a map. The scale of a map sets limits on what is represented and what is omitted. The scale of a map is also used to determine the distance that one unit on the map surface represents on the actual ground surface. As an example:

$$\frac{map\ distance}{ground\ distance} = \frac{1}{500} = 1{:}5000$$

One way to remember the differences between the terms **broad scale** and **fine scale** (as used by landscape ecologists, see Chapter 1, Scale and Hierarchy Theory) and **large scale** and **small scale** (used by geographers in reference to maps) is that to a geographer, a map at a scale of 1:5000 is a larger scale map than a 1:24,000 map because 1 is a larger portion of 5000 than of 24,000 (Monmonier, 1996).

Question 2.1. Using the terminology of a landscape ecologist, is a map rendered at a scale of 1:5000 a fine-scale map or broad-scale map relative to a map at a scale of 1:24,000?

A **planimetric map** such as a road atlas shows only horizontal (two-dimensional) information, while a **topographic map** shows the elevation of objects. Differences in topography are often shown using contour lines (Figure 2.1). The slope between any two points can be determined from the contour lines of a topographic map and can be calculated in a very general form as follows (Ciciarelli, 1991):

$$G = D/H$$

where G = the slope gradient between two points, D = the difference in elevation between two points, and H = the horizontal distance between two points.

Question 2.2. Interpreting the contour lines, qualitatively identify the area that *appears* to be the steepest, as well as the area that appears to be the flattest, on the topographic map for your area. Then, starting with the steeper area, quantitatively determine the slope using the previous formula. Repeat for the flattest area, using the same length line (H) as you used for the steep

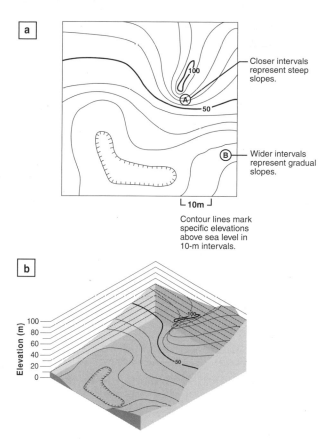

Closer intervals represent steep slopes.

Wider intervals represent gradual slopes.

⌞ 10m ⌟

Contour lines mark specific elevations above sea level in 10-m intervals.

Elevation (m)

FIGURE 2.1
Representations of topographic relief. Each line represents land at a particular elevation above sea level (in this case, in 10-m intervals). (a) Schematic of topographic features as depicted on a topographic map. Steep gradients are represented by lines close together (area A); while contour lines far apart represent gradually sloping areas (area B). Ridge tops are shown as closed loops, while depressions are shown as crosshatched lines. (b) A three-dimensional representation of the topographic map above.

area. Do your quantitative slope measurements confirm your qualitative map interpretation?

The first topographic map was made in 1879 (U.S. Geological Surgery, 1998). Today, many topographic maps are based on information from aerial photographs. Topographic maps from the U.S. Geological Survey (USGS) at a scale of 1:24,000 are referred to as the 7.5 Minute Quadrangle Series. On these maps, 1 inch represents 2000 feet. Topographic maps can be purchased at low cost from most local USGS offices and are now also available online. Different maps may be available outside the United States.

Aerial Photographs

The earliest known aerial photograph was taken from a balloon over a village in France in 1858 (Lillesand and Kiefer, 1994). Taken from aircraft today, aerial photographs can be produced at a variety of scales depending on the altitude of the aircraft and attributes of the camera. When examining your photographs, note that the coverage of an area often overlaps in adjacent photographs.

Measuring *exact* distances on aerial photographs can be problematic for a variety of reasons. **Relief displacement** occurs because low-lying areas are in fact farther from the camera lens and appear smaller in size than areas of higher elevation. For example, a 50-hectare field would look somewhat smaller in a low-lying area and somewhat larger in an area of high elevation. This effect is most apparent in areas of very mountainous terrain. **Tilt displacement** can occur if the camera lens (or more precisely, the optical axis of the camera) is not exactly perpendicular to the ground surface when a photo is taken, but rather is at an angle. This will cause farther objects to appear smaller than closer objects even if they're the same size (Warner et al., 1996). **Orthophotos** are aerial photographs that have been "orthorectified"—that is, errors due to various types of distortion have been corrected.

Question 2.3. Given an aerial photo with no information on the scale of the photograph, how would you determine the scale? *HINT*: One way to approach this is to consider some popular outdoor sports, or see Ciciarelli (1991; 61).

Satellite Imagery

Just as its name implies, **remote sensing** involves the capture of images from some remote distance. Remote sensing can provide information on shape, color, position, temperature, moisture content, and the "health" of vegetation (Wilkie and Finn, 1996). This is often accomplished using satellite imagery; however, aerial photographs, acoustic sounding methods, and radar are also examples of remote sensing. Some commonly used satellites for the collection of spatial data are the Landsat and SPOT satellites. Some considerations for selecting satellite versus aircraft imagery are that aircraft fly at a lower altitude and thus generally collect data at a finer resolution, while satellites cover a greater extent. Wilkie and Finn (1996: 53–60) provide a detailed discussion of the costs and benefits of aircraft versus satellite data. However, the technology associated with satellite remote sensing is rapidly advancing and changing, and data are becoming available at ever-increasing levels of resolution.

Most satellite remote sensing is based on detecting the way surfaces reflect and absorb visible and infrared radiation, a subset of the **electromagnetic spectrum** (Figure 2.2). The percentage of incident light of particular wavelengths that is reflected by an object is referred to as **spectral reflectance** (the total quantity of energy reflected is termed **radiance**). Remote sensors detect the radiance associated with a given pixel and then this information is often converted to the spectral reflectance. Different cover types will each have their own spectral reflectance. For example, chlorophyll in vegetation primarily absorbs radiation in the blue and red wavelengths and reflects radiation in the green wavelengths. Thus, a pixel can be **classified** as deciduous forest, water, or barren soil, for example, based on its spectral reflectance.

Imagery consists of certain **bands**—this refers to the wavelengths of light that are detected by the satellite's sensors. For example, Band 1 in a Landsat-TM satellite detects wavelengths from 0.45 to .52 micrometers (or blue wave-

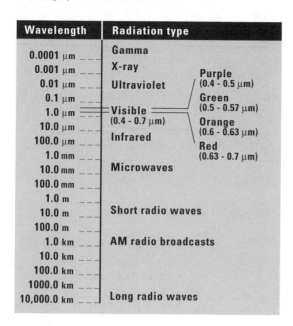

FIGURE 2.2
The electromagnetic spectrum

lengths). Remote sensing satellites can also detect wavelengths outside the spectrum of visible light using infrared bands, allowing the analysis of spatial patterns otherwise invisible to the naked eye. For example, infrared sensors are particularly useful for distinguishing healthy and stressed vegetation and delineating water bodies.

Satellite imagery of your area will be handed out in class. You may be given either classified or unclassified imagery. If provided with both, it is important that you "sample" the unclassified image first, before you see the categorizations used in the classified image.

GIS Data

If GIS data are available for your site, they were likely created using some of the data sources you have just examined. Try to determine which other data source matches your GIS data most closely. Chapter 3 (Introduction to GIS) provides a basic introduction to the components of a GIS as well as some experience using one.

WRITE-UP

Your assignment includes five main parts:

1. Introduction—Briefly describe the objectives of the exercise and how well they were met.
2. Methods
   (a) Discuss your rationale behind transect placement.

(b) Clearly identify the spatial extent, minimum mapping unit, and classification scheme you used.

(c) Discuss any other decisions you made during data collection or analysis that are important for the interpretation of your data.

3. Results—Include your completed **Summary Data Sheet** (Table 2.1 on the CD).

4. Discussion—Address the following questions in your discussion:
   (a) Were there consistent differences in the information obtained from the different data sources?
   (b) How different and/or similar were the results obtained by the different methods?
   (c) What explains the differences and/or similarities in your summary statistics?
   (d) How well did your sampling capture the land-cover types at your site?
   (e) How well do your data portray the fragmentation and connectivity of your site?
   (f) From your experience, discuss the apparent utility of each data source. Are particular types of research questions best suited to particular data sources? Which land-cover types are best observed using the different data sources?

5. Appendix—Attach a copy of your raw field notes.

BIBLIOGRAPHY

Note. An asterisk preceding the entry indicates that it is a suggested reading.

CICIARELLI, J. A. 1991. *A Practical Guide to Aerial Photography with an Introduction to Surveying.* Van Nostrand Reinhold, New York.

* LILLESAND, T. M., AND R. W. KIEFER. 1994. *Remote Sensing and Image Interpretation.* John Wiley & Sons, New York. A widely used, detailed source of information on aerial photography, satellite image sources, and processing.

* MONMONIER, M. 1996. *How to Lie with Maps.* University of Chicago Press, Chicago. Very readable and accessible, includes interesting historical cartographic information.

U.S. GEOLOGICAL SURVEY. 1998. Topographic Mapping. U.S. Department of the Interior, Washington, DC.

WARNER, W. S. , R. W. GRAHAM, AND R. E. READ. 1996. *Small Format Aerial Photography.* American Society for Photogrammetry and Remote Sensing, Bethesda, MD.

* WILKIE, D. S., AND J. S. FINN. 1996. *Remote Sensing Imagery for Natural Resources Monitoring: A Guide for First-Time Users.* Columbia University Press, New York. A good introductory text covering basic concepts.

INTRODUCTION TO GEOGRAPHIC INFORMATION SYSTEMS (GIS)

JOSHUA D. GREENBERG, MILES G. LOGSDON, AND JERRY F. FRANKLIN

OBJECTIVES

Geographic Information Systems (GIS) are important tools for viewing broad-scale patterns of spatial data, organizing and integrating information about an area, and analyzing that data to answer questions. A GIS can be used to ask, What is the total area of parks and preserves within 100 meters of a stream or lake? or, How many different landowners own property in a given area? or, What is the total length of roads and highways that intersect a large predator's home range? As such, a GIS has wide-ranging applications for both management and research in landscape ecology. In this lab, you will

1. gain an appreciation for the utility of a Geographic Information System as an important tool of landscape ecology;
2. learn the basic components of a GIS and gain familiarity with some commonly used terminology; and
3. gain hands-on experience using a simplified GIS program to pose and answer questions.

The exercises in this lab use a simple spatial viewing tool called ArcExplorer, produced by Environmental Systems Research Institute, Inc. (ESRI). After being exposed to the fundamentals of understanding and using a GIS, you will use your knowledge to perform some basic analyses on GIS data from the Gifford Pinchot National Forest, located in the state of Washington (USA).

INTRODUCTION

A Geographic Information System is more than a tool to make pretty maps. The basic GIS provides the user with the ability to store, manipulate, and display information about a region. What separates a GIS from a mere map-making program are the data, which are geographically referenced, can come from many sources, and can be manipulated and analyzed in a variety of ways. Thus, a GIS allows the exploration of more sophisticated spatial questions than would be possible with just a map. In addition, along with the actual software, other important parts of a complete GIS are the people and resources required for support (Chrisman, 1997). Most people who work with GIS soon realize that they depend on the computer system administrator (to help them keep the machine running); the data (which needs to be accurate for a particular use); and a host of connected tools, people, and software to get a project completed.

Importance of GIS in Landscape Ecology

In the 1970s the importance of analyzing ecological processes and conducting management at broad spatial scales became very apparent. Natural scientists were finally becoming seriously interested in the effects of spatial pattern on ecological phenomena, including pattern and process at the broader spatial scales represented by the "new" topical area of landscape ecology. Resource managers were increasingly challenged to develop management plans that incorporated large areas and long time periods as well as to place specific projects in a spatial context for their analyses.

Despite the need, the creation and utilization of large spatially explicit data sets for such analyses were very difficult; the necessary tools for creating, storing, and maintaining such data sets simply did not exist. Consequently, scientists were constrained in their selection of research topics, spatial scales, and hypotheses. Managers were similarly limited to what could be accomplished with multiple map-based overlays and presentation of a few static alternatives. Addressing spatially explicit issues was clumsy and laborious.

The advent of GIS has revolutionized landscape ecology and, more generally, both basic and applied spatial analysis. The tools now exist to gather, store, manipulate, analyze, and present large spatially explicit data sets. Today scientists can ask questions that are much more sophisticated and complex in numbers and scale than they could even ten years ago. Indeed, many earlier efforts at spatial analysis are embarrassing in light of the exponential increase in the size of areas and phenomena that can now be surveyed using remote imagery and manipulated using GIS. Forest policy and environmental analysis has similarly entered a new millennium with the new tools and the highly interactive capability they provide. Resource managers, decision makers, and stakeholders can all access and manipulate the same large data sets and explore long-term consequences of various policy and project decisions, limited only by the creativity of the GIS user.

Currently, the GIS industry provides a wide variety of software packages accessible to a wide range of people, from desktop computer users to high-end researchers using networks of powerful computer stations. Next, we examine some general features of GIS data that are relevant to nearly any type of GIS program.

Spatial Data and Attribute Data

A GIS deals with the representation of both spatial data and attribute data. **Spatial data** consist of the location of objects on the surface of the earth, while **attribute data** consist of other types of descriptive information about an object (Figure 3.1). An example of spatial data might be the location of houses or the delineation of a road. This locational information is stored using a series of *x*, *y*-coordinates based on a **map projection** (i.e., a method to transpose data from the earth's curved surface to a flat map with minimal distortion). The associated attribute data, however, does not tell you *where* the object is on the earth, but it contains other important information about the object. An example of attribute data might be the age of a house or the average daily temperature of a stream. While the area of a lake, for example, might at first seem like spatial data, in the context here, it is considered attribute data and is listed in the attribute table with other information (Figure 3.1, polygon A). Although it may be spatial data in a broad sense, it is not *spatially explicit locational* data (i.e., a 4-ha lake could be located anywhere). Being able to link the spatial data with the attribute data is the critical component of a GIS that allows complex analyses to be performed. In addition, since the data are georeferenced to real-world coordinates through the map projection, multiple data sources can be combined and analyzed.

Spatial Data Formats: Vector vs. Raster

Two alternative methods of representing spatial data are the **vector** and the **raster** data structures (Maffini, 1987). Vector data are stored as points, lines, or polygons (Figure 3.2). The spatial information is linked to the attribute

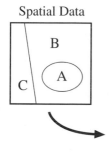

Spatial Data	Attribute Data

Polygon	Class	Area
A	Water	1
B	Forest	4
C	Urban	2

FIGURE 3.1
Spatial data and the associated attribute data for three land classes using vector representation

FIGURE 3.2
Vector representation of spatial data

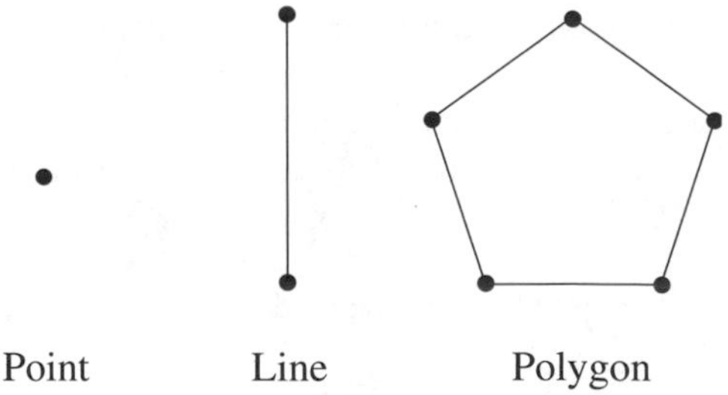

Point   Line   Polygon

data through a common identification (common-id) label. In Figure 3.3, the numbers 1, 2, and 3 are the common-ids that link the spatial objects to the attribute data. Raster data are composed of a regular grid network, usually of square cells (Figure 3.3). Each cell has a value that is used as the common-id to link to the attribute data, regardless of whether all the neighboring cells are the same. A raster data representation is often used for images such as satellite data, or continuous data such as elevation. In addition, spatial simulation models often use a raster data format. The difference between raster data and vector data can usually be discerned visually by identifying the square staircase shapes of edges and lines found in raster data.

Early on in GIS development, the decision to use raster or vector data was often based on the features to be represented and the disk space required by vector and raster formats. A GIS is only as accurate as the original data, and the use of vector or raster data can greatly influence the shape of the resulting landscape features. For example, roads and houses might be better suited for vector lines and points, while topography might be better suited to raster. Given today's faster computers and larger disk storage space, disk space is

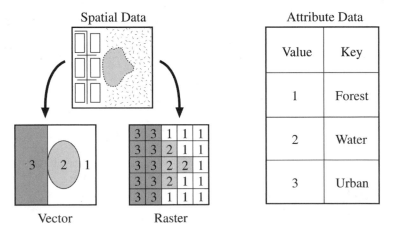

FIGURE 3.3
Raster and vector representation of a landscape with three classes

usually less of a consideration (or limitation) to using either a raster or vector format.

Attribute Data

In addition to knowing the spatial location of an object, equally important is the attribute data linked to an object. Consider the kinds of variables that can be stored as attributes:

1. Nominal variables (those described by a name without a specific order)
2. Ordinal variables (an ordered list of discrete classes)
3. Interval/ratio variables (a sequence of values with a meaningful magnitude between them)

For example, one may wish to record attribute data of vegetation patches such as the name of the dominant tree species (nominal), the age class of forest stands (ordinal), the average temperature in Fahrenheit (interval), or the amount of intercepted rainfall per month (ratio).

An understanding of the differences in these types of variables is necessary to correctly conduct further analyses with attribute data. For example, adding nominal or ordinal class data together could give a numerical result, but it may be meaningless (e.g., adding $1 + 2 = 3$, given that 1 = old growth and 2 = young stands). This is critical since many steps can be involved in GIS analysis, and undetected errors could corrupt the quality of future data.

Linking Spatial and Attribute Data

The combined spatial and attribute data have many names depending on the GIS software or text being used. In this lab, we will refer to the collective spatial and attribute data as a **theme,** corresponding to the terminology used by the software we will use later in the lab. However, words such as *coverage* and *layer* are also commonly used in the same sense. A theme is usually restricted to a particular subject being represented. For example, "streams" might constitute a theme, separate from the "property boundaries" theme (Figure 3.4). In landscape ecology, some common themes are flowing water (streams and rivers), still water (lakes and reservoirs), roads, land use, land cover, elevation, habitat maps, and land ownership. Obviously, themes will vary depending on the initial source and intended use.

GIS Analysis using Boolean Logic

At the simplest level, a GIS helps the practitioner answer two fundamental questions: Where are all of the objects that meet the following conditions? and, What is at this location? (Huxhold, 1991). Before the widespread use of GIS, such analyses were conducted by overlaying transparent maps on a light table and looking for patterns based on the resulting dark areas as well as the light showing through the various map layers (McHarg, 1969). This type of analysis is now automated within a modern GIS using the rules of Boolean logic.

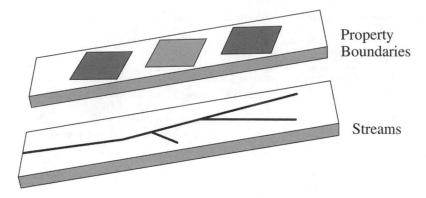

FIGURE 3.4
Example of two different themes (or coverages) for an area

Boolean logic, named after the British mathematician George Boole, deals with the sorting of data through simple combinations of questions and is used to construct query statements. A query statement operates on attribute data and is used to locate the objects that meet the conditions of the Boolean logic. The simple Boolean operators are the familiar **AND, NOT, OR,** and **XOR** operators and are best understood in the form of Venn diagrams (Figure 3.5).

An example of a Boolean question might be to select all urban areas that contain vegetation; the Boolean equation would be urban **AND** vegetation. While this example represents the very simplest query, drawing a picture of your question using Venn diagrams may be necessary for deriving the correct Boolean equation for more complicated queries.

Many other analyses can be conducted with a GIS. Some of the more common spatial functions are measuring the distance between objects, delineating buffer zones (areas that are a fixed distance from an object), and determining

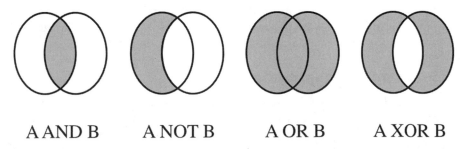

A AND B   A NOT B   A OR B   A XOR B

FIGURE 3.5
Venn diagrams depicting results of applying Boolean logic. Set A is the polygon having attribute A (the circle on the left of each example), and set B is the polygon having attribute B (the circle on the right). The shaded areas are returned as testing "true" to the stated conditions in a query statement.

the spatial arrangement of objects (clustered, evenly spaced, connected). Next, you'll explore some basic commands and analyses using a simple, user-friendly version of a GIS.

The lab that follows will use a spatial data visualization program called Arc-Explorer (A/E) produced by Environmental Systems Research Institute, Inc. (ESRI). Although similar to the popular ArcView and ArcInfo, it is not a complete Geographic Information System. However, ESRI offers this software free of charge, so you can use it in lieu of other, more expensive programs. While the lab exercise directions are oriented to using A/E, the spatial data you'll be using are in Arc/Info coverage format and can also be viewed in Arc/Info or Arc/View. In this lab you will learn how to open a project, display data themes, alter the appearance of data, and perform some simple queries by using our provided A/E directions. For a more complete set of directions, check the A/E help menu.

The exercise uses GIS data from the Gifford Pinchot National Forest in southwest Washington state. An extensive fire in 1903, the "Yacolt burn," left a mosaic of burned and unburned patches throughout the area. Areas not burned in the Yacolt fire are now classified as old-growth forest (over 200 years old), while areas that did burn are classified as either young forest (41–80 years old) or mature forest (81–200 years old). Since the Yacolt fire, some timber cutting has occurred, primarily in the older forest stands. The classification scheme focuses on dividing conifers into age classes, but also includes a class for hardwoods (usually alder, which have traditionally had little timber value) and bare areas (usually rock outcrops or avalanche shoots). More information is available in the metadata file on the CD (called **metadata.pdf**).

EXERCISE
Installing ArcExplorer and Copying Associated Files

To install ArcExplorer, double-click on the file **aeclient.exe** provided on the CD and follow the directions. Then, be sure to copy the *entire* **GISData** folder to a directory in your computer. This will maximize the speed of the program and allow you to save changes to the project. Better yet, copy the whole directory for Chapter 3 to your computer before you begin.

BASIC INSTRUCTIONS FOR ARC/EXPLORER

Opening a Data Set

1. Start ArcExplorer by double-clicking on its icon. (The default settings for installation would place an ArcExplorer icon on your desktop, or look under C:\Program Files\ESRI.)
2. Select **File**, then **Open Project** using the pull-down menu.
3. Then select **intro.AEP** from the GISData directory you copied to your computer.

Intro.AEP is a project, or a file that contains pointers that tell A/E where the data exist and how to display them—but *does not actually contain the spatial data!* Therefore, if you copy the **intro.AEP** file to another location, the coverages will not also be copied. Thus, be sure to keep all .AEP files and the accompanying files together in the GISData folder.

☑ Displaying Themes

The **intro.AEP** project has a legend editor on the left (Figure 3.6) and a map on the right. The legend editor lists the themes that can be displayed. When a theme has a check, it is displayed in the map window on the right. Check the appropriate boxes to display the **10m_contours, streams,** and **roads** themes. Note the difference between a theme being checked for display and a theme that is made active. An **active theme** is the one selected for analysis. By clicking on the name of the theme in the legend (not the check mark), the theme will become active as shown by a shadow box around the theme. This means that any theme functions and queries you now perform will be made on that active theme only. The vegetation cover theme is activated in Figure 3.6.

A/E always draws themes in reverse order beginning with the theme listed at the bottom of the list and then draws each subsequent theme on top. To see how this can affect your final map, try drawing the vegetation cover map last (at the top) and notice what happens to all the line data (roads and streams). To do this, select the vegetation theme (make it active) and then drag and drop it to the top position. Move the vegetation theme back to the bottom before proceeding with the rest of this lab.

Browsing Themes

Several functions allow you to move around the map region. Familiarize yourself with the following buttons:

FIGURE 3.6
Legend editor

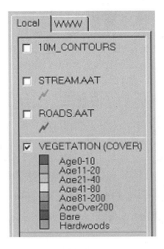

Zooms in / Zooms out when button is clicked and then mouse is used to click and drag over a portion of the map window.

Allows you to pan the map (useful when zoomed in on a small area). Click on the hand button, then click and drag the mouse over the map window.

Moves to the extent of all themes / Moves to the extent of the active theme.

Question 1. Use the zoom tools to determine if the vegetation theme consists of vector or raster data, and explain your decision.

Theme Legends

Each theme contains a key in the legend. The categories shown in the key represent one category (or **field**) of the **attribute data.** The list of categories in the key for the vegetation theme represents all the discrete values for that theme. The number ranges refer to the **ages** of the conifer forests, **Bare** refers to areas of bare ground, and **Hardwoods** are the deciduous forests (in Washington state these are often associated with stream areas)

Altering Theme Legends

To alter a theme's appearance, try the following:

1. Ensure that the **Vegetation** theme is active.
2. Click the theme menu button shown above (or double-click on the legend for that theme) to open the **Theme Properties** window (Figure 3.7)
3. In the Theme Properties window, under **Discrete Values and Symbols,** select the category you wish to change by clicking on the colored box next to that category (Figure 3.8)
4. Then, in the **Symbol Properties** window, click inside or pull down the box to the right of **Color, Style, Size, or Outline Color** to alter each of these.
5. Click **OK** then **Apply** to activate the change.

Question 2. Make a rough visual estimate of the percentage of bare ground (the **Bare** category) on the image and record your answer. Then, change the

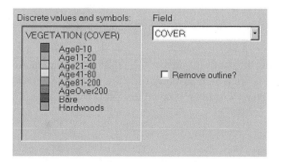

FIGURE 3.7
Theme properties

FIGURE 3.8
Location to place cursor in order to change the color
of a class

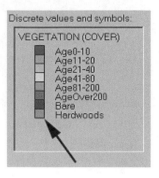

color of the bare ground category to some very bright contrasting color. Does
this change the way you visually interpret the map—that is, does bare ground
seem more or less abundant in the new color?

The visual appearance of maps is an important part of qualitative landscape analysis, and the way data are displayed can influence the way the
viewer interprets a map. Using a brighter, more obvious color for a map
category can make it appear more abundant than it really is. For examples
of other subtleties in map making, see *How to Lie with Maps* (Monmonier,
1996) listed as a suggested reading in the bibliography.

There are other ways to alter the appearance of a theme, such as selecting different classification schemes. Consult the A/E help files for more
information on these features.

 Printing

The Print button is the quickest way to print your map. Unfortunately, A/E
does not have very sophisticated map layout options like some of the more
advanced GIS programs. If you want to modify a map layout in A/E, try
making a screen capture (Ctrl-Print Screen) and opening the file in a graphics program. You may find that printing a map is the best way to illustrate
a point you are making. Check with your teacher or computer administrator about selecting a printer on your computer system to use for printing.

QUERYING A THEME

 Distance Tool

The Distance tool calculates the distance between two spots on a map via
the following steps:

1. Pull down the small arrow menu to the right of the measuring tool
   and select the units in which you would like your measurement displayed. *NOTE*: The kilometer distance is incorrect due to a bug in
   the software.
2. Select the Distance Measuring Tool icon.
3. Place the cursor on the map.

4. Click and drag (hold the click while dragging) to start measuring, and then release to stop measuring.

5. The distance between the points will be shown in the upper left corner of the map area.

6. Double-click the mouse anywhere on the map to reset the distance to zero.

Question 3. What is the width (in meters) of the extent for the entire vegetation theme?

 Find Tool

The Find tool searches a theme to locate text in the attribute table associated with a theme. Once it locates the text, it displays all points, lines, or polygons that meet the specifications.

1. Click the Find tool, which opens the **Find Features** window.

2. Enter the text you want to search for (e.g., Bare). Be aware that the Find tool is case sensitive.

3. Select a search type.

4. Choose the theme to search in (e.g., vegetation).

5. Click Find. The results are displayed in the lower box labeled **Pick a Feature**.

6. After you pick a feature, you can then highlight or zoom to the selected items.

Alternative Method

Click on the colored box of a category in the legend. You will notice that all the polygons of that category are selected and highlighted in yellow in the map window, and a box in the lower left corner of the map window indicates which category is selected.

To deselect, use the eraser 🖉 button.

ⓘ Identify Tool

The Identify tool retrieves attribute data from the selected point, line, or polygon.

1. Activate the theme that contains the feature you wish to query.

2. Click the Identify button

3. Click on a point, line, or polygon on the map. Be sure it occurs in the active theme!

4. A window will appear displaying the attribute information for the selected feature.

NOTE: The units for perimeter and area are in meters and meters squared for the polygons in this particular map

Question 4. What is the perimeter and area of the large polygon, class Age41–80, in the northwest corner of the map?

 Query Builder

A more advanced method of making a data query is to use the A/E Query Builder. This will open a window that allows you to use an equation to select records based on the attributes of the active theme. By using Boolean logic equations in the query editor, you can create some complex queries from the data set and run simple statistics on the results.

1. Ensure that the vegetation theme is active.

2. Click the Query Builder button from the menu bar.

3. On the left of the Query Builder window, click on a field listed under "Select a field." (e.g., cover).

4. Remembering your knowledge of Boolean logic, select the query operation to perform from the operator buttons shown in the center of the window.

5. Type in the value or word to query or select a sample value from the displayed list. The Query Builder is case sensitive so use capitals when needed. Also, when typing in text, enclose the text in single quotation marks.

6. Click Execute.

For example, if cover is selected, you can find all the Bare polygons by creating the following equation (Figure 3.9) (you can type the equation, or click the Option buttons in the windows) and then clicking the Execute button.

FIGURE 3.9
Equation editor with an equation to select all Bare class polygons

7. The area lower in the Query window will show you the results of your selection.

8. By clicking the **Show All Attributes** box, you can view complete information for all the polygons that matched the query (in this case, all the bare ground polygons).

9. Click on the **Highlight Results** button and the query selection is shown in yellow on the map.

10. Clicking on the item headings (such as Area) will reorder the results in ascending or descending order.

11. To calculate some basic summary statistics, press the **Statistics** button and choose the Area item. In this case, be sure to check the **Use Query Results** box also. The bottom window will show you statistics on the area of the bare ground polygons.

X This button clears the equation selection to allow you to perform a new query.

Question 5. Using the Query Builder (use a calculator to find the percent of total extent), fill in Table 3.1. You can print out a copy of Table 3.1 from the CD.

Question 6. How much old-growth forest (over 200 years old) exists that occurs in stands larger than 100,000 m²? Include the Boolean formula you used.

Saving Queries to a Spreadsheet

After making a query, your results can be imported easily into a spreadsheet program such as Microsoft Excel to conduct further analyses, perform statistical tests, or make graphs and tables. This process assumes that you have some previous experience with a spreadsheet or graphing program. To export a query:

1. After performing a query, click on the Save button in the Query Builder window. In order to export the complete database, make a query that includes all the polygons (i.e., area > 0), and then export that query.

2. Specify the directory in which to save the file.

3. In a spreadsheet program such as Excel, import the file as a delimited file using the "|" as the delimiter.

TABLE 3.1
DATA TABLE FOR QUESTION 5

Category	Area	Percent of Total Extent
Age 81–200		
Age Over 200		
Bare		

Optional Question 7. This question assumes that you know how to make graphs using spreadsheet software. Make scatter plots showing the area and perimeter of the AgeOver200 and Bare vegetation categories. Describe any differences or similarities that you see in the plots and explain why those differences or similarities exist. (Your plots should be laid out like the examples in Figure 3.10)

Not all of the functions of ArcExplorer have been covered in this lab; additional information can be found in the A/E help files.

Advanced Exercises

Using what you have just learned, conduct the following series of analyses. *NOTE:* We haven't given you such detailed instructions this time! If you get stuck, refer back to the ArcExplorer directions we provided.

Adding a Theme

If you want to add a theme to your project, or create a new project with your own specified themes, click on the **Add Theme** button and use the menu to select the themes that you would like to add. Notice that you can also add themes from data sets that exist on the Internet, or other Arc coverages that may be available to you on your local drive or network.

Add the **StreamVeg** theme to your project. This theme contains information about the vegetation that lies within a buffer zone around a stream network. A buffer zone consists of the area at a fixed distance from a feature. Often buffer zones are used around streams to determine the riparian areas, which have unique flora and fauna. The cover data is the same as for the vegetation theme; however, there is also a **class** item field. The categories of **class** data are either "riparian" or "upland," identifying the areas that are in the buffer versus areas that are outside the buffer, respectively.

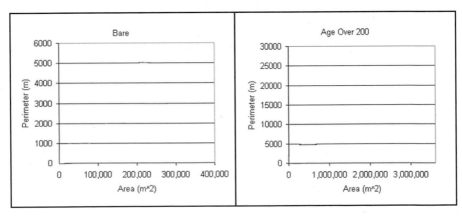

FIGURE 3.10
Example plots for *Optional Question 7*

Question 8. Determine the buffer distance used in creating the StreamVeg theme.

Question 9. Calculate the area of old-growth, hardwoods, and bare ground within the riparian buffer.

Question 10. How does the relative area occupied by the different vegetation types in the riparian areas (measured in the previous question) differ from what you would expect to find if those cover types were randomly distributed throughout the entire landscape? Explain why the difference might exist and how you determined it.

Question 11. What issues might a land manager be concerned about for a species that can only move in riparian regions? Think about the connectivity of the stream buffer area and how it relates to elevation using the 10-meter contour and roads themes.

Question 12. Metadata are essentially data about data. Metadata contain much more detailed information about the data being used in a GIS, including how and when they were produced, the name of the creator, and the purpose of the original maps. What are some metadata that you might find helpful for the data set used in this lab? Examine the file **metadata.pdf**, which can be found on the CD in the directory for this chapter. It can be viewed using the freely available Adobe Acrobat software.

BIBLIOGRAPHY

Note. An asterisk preceding the entry indicates that it is a suggested reading.

*ARONOFF, STANLEY. 1989. *Geographic Information Systems: A Management Perspective.* WDL Publications, Ottawa. A good introduction to GIS, and also covers some remote sensing. Although this book is a little outdated, it still provides an easy-to-read presentation on GIS.

CHRISMAN, N. R. 1997. *Exploring Geographic Information Systems.* John Wiley and Sons, New York.

*HAINES-YOUNG, R., D. R. GREEN, AND S. COUSINS. 1993. *Landscape Ecology and Geographic Information Systems.* Taylor & Francis, New York. Examples of using GIS for landscape analysis in a variety of ways.

HUXHOLD, WILLIAM E. 1991. *An Introduction to Urban Geographic Information Systems.* Oxford University Press, Oxford, UK.

MAFFINI, G. 1987. Raster versus vector data encoding and handling: A commentary. *Photogrammetric Engineering and Remote Sensing* 53(10):1397–1398.

MCHARG, I. L. 1969. *Design with Nature.* Natural History Press-Doubleday, Garden City, New York.

*MONMONIER, MARK S. 1996. *How to Lie with Maps.* University of Chicago Press, Chicago. An interesting look at cartography in a fun, easy-to-read format.

MODELS AND CAUSES OF LANDSCAPE PATTERN

Spatial models are important tools for exploring landscape change. In Chapter 4, Introduction to Markov Models, students calculate transition probabilities for the analysis of landscape change. Markov models were one of the earliest types of models used to examine landscape change, and many more complicated models of landscape change are based on the basic principles of Markov models. Chapter 5, Simulating Changes in Landscape Pattern, presents an application of simulation modeling to forest harvest management, allowing students to conduct their own simulations using the Harvest Lite model. This lab also examines the assumptions and limitations involved in simulation modeling. While Chapter 4 examines changes over the past several decades, and Chapter 5 presents an example of projecting much longer into the future, Chapter 6 illustrates the challenges inherent in examining the causes of landscape change over even longer (historical) time scales. Chapter 6, Creating Landscape Pattern, is a case study examining spatial data from the state of Michigan (USA) to examine the diverse, long-term causes of landscape heterogeneity and the interaction of human-induced and natural causes of spatial heterogeneity.

INTRODUCTION TO MARKOV MODELS

DEAN L. URBAN AND DAVID O. WALLIN

OBJECTIVES

Models of landscape change are important tools for understanding the forces that shape landscapes. One motivation for modeling is to examine the implications of extrapolating short-term landscape dynamics over the longer term. This extrapolation of the status quo can serve as a frame of reference against which to assess alternative management scenarios or test hypotheses. There are a spectrum of ways to consider landscape change, ranging from simple and readily interpretable, to more realistic and less tractable. The goals of this lab are to

1. provide an introduction to the mathematics of simple Markov models;
2. enable students to build a simple model of land-use change based on transition probabilities;
3. explore the process of model creation, verification, and validation; and
4. encourage creative speculation as to how Markov models might be extended to incorporate more complex and realistic mechanisms of and constraints on landscape change.

In this exercise, you will build a simple model of landscape change, evaluate it, and use it as a point of departure to consider more realistic (but more complicated) models. Raster maps of Pacific Northwest forests are compared over three time periods to summarize the rates of transition between cover types.

A simple model of landscape change is built from these transition probabilities. This model is then projected forward in time to verify and validate the model. In order to complete this lab, you will need a PC, a calculator, the program **markov.exe**, the landscape images (**pnw72.gif, pnw85.gif, pnw91.gif,** and **samp200.gif**), and the accompanying data file (**samp200.dat**) located in the directory for this lab on the CD-ROM.

INTRODUCTION

Perhaps the most fundamental observation of landscape change arises from measurements of the state of a landscape at two time periods. For example, we might have land-cover maps classified from satellite images obtained for two dates ten years apart and note that some of the cells (or pixels) changed cover type over that time interval.

   One way to summarize landscape change is simply to tally all the instances, on a cell-by-cell basis, in which a cell changed cover types over that time interval. A concise way of summarizing these tallies is a **raw tally matrix,** which for m cover types is an $m \times m$ matrix. The elements, n_{ij}, of the tally matrix tally the number of cells that changed from type i to type j over a time interval. A raw tally matrix is often converted into proportions by dividing each of the elements by the row total to generate a **transition matrix** P. The elements, p_{ij}, of the transition matrix P summarize the proportion of cells of each cover type that changed into each other cover type during that time interval. The diagonal elements of this matrix, p_{ii}, are the proportions of cells that did not change.

   While a variety of approaches to modeling landscape change exist (see Weinstein and Shugart, 1983; Baker, 1989; Sklar and Costanza, 1991 for reviews), many of these begin with a tally matrix or the transition matrix P. Here, you will examine the simplest of such models based on a transition matrix. This simple model will serve as a point of departure for the contemplation of more realistic, but also more complicated models.

Markov Models

A **first-order Markov model** (Usher, 1992) assumes that to predict the state of the system at time $t + 1$, one need only know the state of the system at time t. The heart of a Markov model is the **transition matrix P,** which summarizes the probability that a cell in cover type i will change to cover type j during a single time step. The time step is the interval over which the data were observed to change (i.e., the time interval of the two maps).

   Markov models, while simple, have a number of appealing properties. In particular, they can be solved by iteration to project the state of the system. Writing the state of the system as a vector,

$$x_t = [x_1 \; x_2 \; x_3 \; \ldots \;] \qquad (1)$$

where x_i is the proportion of cells in type i at time t, a Markov model is projected:

$$x_{t+1} = x_t P \tag{2}$$

that is, the state vector postmultiplied by the transition matrix. The next projection for time $t + 2$ is continued:

$$x_{t+2} = x_{t+1}P = x_t PP = x_t P^2 \tag{3}$$

and in general, the state of the system at time $t = t + k$ is given by:

$$x_{t+k} = x_t P^k \tag{4}$$

where x_t is the initial condition of the map. Thus, the model can be projected into the future simply by iterating through the matrix operation (see Exercise 2 for details on how to do this manually).

The steady-state or equilibrium state of the system is given by the eigenvector of the transition matrix; thus, there is a closed-form solution to the model. Recall, the **eigenvector** of the matrix is defined such that the matrix multiplied by the eigenvector yields the vector again:

$$\tilde{x} = \tilde{x}P \tag{5}$$

That is, the system does not change once it reaches this state. There are some computational tricks for estimating steady-state solutions (Usher, 1992), or you could use a math package (e.g., Mathematica, MatLab) to do this. But for simple models, the solution often converges rapidly, and you can estimate the solution simply by projecting the model a few times.

Graphical Representation

The model implied by the transition matrix P can also be represented as a graph (a "box-and-arrow" diagram). An example with three cover types could be illustrated as in Figure 4.1. Casual inspection of the graph reveals the direction of flow in the system and suggests a succession from type 1 through type 2 to type 3, with some recycling (possibly a disturbance) to the initial cover type.

Model Projection

To explore a Markov model, it is initialized with a state vector and then projected for one or more time steps. The vector of cover types produced at each iteration is the prediction of overall landscape composition for that time step. In the following exercises, we will show you how this is accomplished.

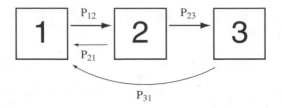

FIGURE 4.1
A schematic box-and-arrow diagram of a transition matrix P with three cover types. The thickness of the arrows indicates the magnitude of the transition rates between the different cover types (the arrows for self-replacement are not shown). The diagram shows flow from cover type 1 to 2 to 3, with some recycling to previous cover types via disturbance.

Modeling Landscape Change in the Pacific Northwest, USA

Much debate over the management of Pacific Northwestern (PNW) forests occurred in the early 1990s. The debate centered on the effects of intensive logging on old-growth forests and on old-growth-dependent wildlife species such as the northern spotted owl (*Strix occidentalis*) and the marbled murrelet (*Brachyramphus marmoratus*) (Hansen et al., 1991; Ruggiero et al., 1991), primarily on U.S. Forest Service land. Since most of the old-growth on private lands was already gone, most of the old growth harvested in the 1980s came from these federal lands (Harris, 1984; Robbins, 1988). By the mid-1990s, harvests were reduced by over 80% relative to the peak harvests of the late 1980s (FEMAT, 1993; USDA Forest Service and USDI Bureau of Land Management, 1994; Marcot and Thomas, 1997). A central questions during this debate was, How long can current rates of harvest be sustained before the old growth is virtually gone?

Study Area

The Oregon Cascades were at the center of this debate over the management of PNW forests. The study area is on federally managed lands where timber harvesting has been conducted using a dispersed ("staggered setting") system of 10- to 20-hectare clear-cut patches. The rate and pattern of these disturbances is somewhat different than those on private lands (Spies et al., 1994) and is quite different from disturbances generated by wildfire during the pre-settlement era (Wallin et al., 1996b).

Spatial data for this area were derived from Landsat Thematic Mapper data using methods outlined in Cohen et al. (1995, 1998) and Wallin et al. (1996a). Forest cover was classified into six approximate age classes (Table 4.1). Images for three time periods are included here: 1972, 1984 and 1991. The images for the different time periods (**pnw72.gif, pnw84.gif,** and **pnw91.gif**) can be examined by using a web browser (use **File,** then **Open** on your web browser's pull-down menu). Yellow areas denote young stands, successively darker greens are older forest, and brown areas are recent clear-cuts. The gray

TABLE 4.1
DEFINITION OF COVER TYPES IN THE PACIFIC NORTHWEST-
ERN FOREST LANDSCAPE.

Class	Age (yr)	Cover Type
0	Background	(Nonforest)
1	0–20	Recent clear-cut
2	21–40	Early seral
3	41–80	Mid-seral
4	81–170	Mature
5	>170	Old growth

spot in the images is rock. Each image is 500×500 cells (15,625 ha) with a cell size of 25 meters.

The file **samp200.gif** shows 200 random locations on the 1972 image. The cover type at each of these locations was tallied at time 1 (1972), 2 (1984), and 3 (1991) on the three maps. This information is tallied in columns of a **primary data matrix** and can be viewed by opening the text file **samp200.dat.**

Next, you will use the cover type data from the 200 sample points to build your own Markov model. In the following exercises you will perform three main steps: model development, model verification, and model validation.

EXERCISE 1
Model Development

Here, you will use the data from the 200 sample points on the PNW images to calculate transition probabilities for a Markov model. The transition probabilities will be based on landscape change from 1971 to 1984.

1. From the primary data matrix (**samp200.dat**), construct a raw tally matrix that summarizes the number of the 200 cells that underwent a transition from type i to type j during the time period t_1 (1972) to t_2 (1984). Recall that each element, n_{ij}, in the tally matrix is the number of times a cell changed from type i to type j during the time interval. Print out and then enter your results in Table 4.2 on the CD.

2. Divide each element in the raw tally matrix by its row total to yield a matrix of transition probabilities p_{ij}, the probability (or rate) of change from type i to j. These probabilities are on a 12-year time step (1972–1984). Enter your results in Table 4.3 from the CD. Calculate your results to five decimal places.

3. Next, convert the transition matrix to an annual time step. This is partly cosmetic (transient dynamics will look smoother), but will also make it possible to reconcile the 12-year time step of the first period with the 7-year time step of 1984 to 1991. To convert the transition probability matrix P to an annual time step, do the following:

   (a) Divide each of the off-diagonal elements p_{ij}, $i \neq j$, by 12.

   (b) Adjust the diagonal elements p_{ii}, to be $1.0 - \sum_{j} p_{ij}$. In other words, all rows must sum to 1.0. Enter these results in Table 4.4 from the CD.

   In this matrix (Table 4.4), the off-diagonal rates are now annual transitions (probabilities). The diagonal elements are now larger than in Table 4.3 because on an annual time step fewer of the cells actually change, and again, the rows of the matrix must sum to 1.0.

5. Now use Table 4.5 to summarize the state of the map at *all* time steps. The state of the map is defined by a row vector, the elements of which are the proportion of cells in each cover type in each of the maps. Construct three summary vectors from the primary data matrix (**samp200.dat**). To construct these vectors, simply sum up all of the cells in each state in each of the three years of sampling. Convert these numbers to proportions by dividing each element of the table by its row sum. (These vectors can also be derived from the transition matrices. How?)

With these state vector tallies and the transition probability matrix P, you have constructed a simple model of landscape change. All that remains is to evaluate the model. The 1972 data will be used as initial conditions for the model. The second time period (1984) will be used to verify the model. The third time period (1991) will be reserved to validate the model.

EXERCISE 2
Model Verification: Matrix Projection by Hand

Model verification consists of testing the model against the data used to construct it (Haefner, 1996). In this case, the test is of the model projection from 1972 to 1984, compared to the actual data from 1984. Because the model was built from these data, this is *not* an independent test of the model; the model *should*, in fact, match these data. You will verify the model by initializing it with 1972 data and projecting it to 1984, both by hand (here) and by using a computer program (Markov) in Exercise 3. Do the first matrix projection (from 1972 to 1984) by hand following the example below. To do this in one iteration, use the transition probablities in Table 4.3, which are for a single 12-year time step.

This example shows a generic matrix projection using only three cover types. The first step entails multiplying the transition probability matrix P by the state-vector.

$$[x_1 \ x_2 \ x_3] \cdot \begin{bmatrix} p_{11} & p_{12} & p_{13} \\ p_{21} & p_{22} & p_{23} \\ p_{31} & p_{32} & p_{33} \end{bmatrix} =$$

$$[x_1 p_{11} + x_2 p_{21} + x_3 p_{31}, \ x_1 p_{12} + x_2 p_{22} + x_3 p_{32}, \ x_1 p_{13} + x_2 p_{23} + x_3 p_{33}]$$

Note the subscripts to ensure that the proper elements are being used: the inner subscripts must always match (i.e., the subscript on x must match the first subscript on p).

An example using actual data with only three cover types, would look like this.

(a) Suppose the following transition probability matrix (P) is used:

$$P = \begin{bmatrix} 0.90 & 0.10 & 0.00 \\ 0.10 & 0.80 & 0.10 \\ 0.10 & 0.00 & 0.90 \end{bmatrix}$$

(b) Next, suppose all of the landscape is assigned to cover type 1 to begin:

$$x_0 = [1.00 \quad 0.00 \quad 0.00]$$

(c) The first projection, to t_1 is:

$$[1.0 \cdot 0.9 + 0.0 \cdot 0.1 + 0.0 \cdot 0.1, \ 1.0 \cdot 0.1 + 0.0 \cdot 0.8 + \\ 0.0 \cdot 0.0, \ 1.0 \cdot 0.0 + 0.0 \cdot 0.1 + 0.0 \cdot 0.9] = [0.9 \quad 0.1 \quad 0.0]$$

The second projection to t_2 uses the resulting vector [0.9 0.1 0.0] and the original transition probability matrix, and produces:

$$[0.82 \quad 0.17 \quad 0.01]$$

You should verify this by hand. Also, note that these same data are used in the demo data file supplied with the lab (**demo.dat**).

EXERCISE 3
Matrix Projection Using the Program MARKOV

Not surprisingly, matrix projections are often accomplished with the use of a computer program. Here, you will use a simple fortran program called MARKOV. The program (**markov.exe**) can be run from a DOS prompt or by double-clicking directly on its icon. Demo data are available (**demo.dat**).

MODEL INPUT
MARKOV expects to read a user-provided ASCII data file containing the transition matrix and a vector of initial conditions. For these exercises (with five cover types) the data files must be formatted as follows:

Rows 1–5: the elements of the transition matrix (use Table 4.4 for an annual time step)

Row 6: the initial conditions (the 1972 row from Table 4.5)

The data values themselves can be delimited by spaces, a comma, or tabs, and you should use enough significant digits to avoid round-off error (say, five decimal places).

MODEL OUTPUT

The program will report either the time step at which the solution converged to steady-state, or that the model did not converge during the simulation. In the latter case, you need to rerun the model for a longer time so that it has time to converge. The output written by MARKOV consists of one line per time step: reporting time step (in column 1), followed the proportion of the landscape in each cover type at that time (in this case, columns 2–6). The output file from MARKOV is formatted so that it can be imported directly into a spreadsheet or graphics package.

A session with MARKOV is as follows:

```
Project a markov Model?
  Name of file with input data?
  And name of file for output data?
  Number of patch types in model?
  And number of timesteps to project?
Model failed to converge in 100 timesteps
```

When asked for the number of patch types, this refers to the number of cover types in the model.

Using the the program MARKOV, repeat your model projection from 1972 to 1984 using the following steps:

1. Make an input file (a text file) that includes your matrix and vector data. Format your input data as explained in the Model Input section. *NOTE*: You can refer to the file **demo.dat** for guidance, but remember that these data are for the example with only three cover types.

2. Run the program MARKOV for 12 time steps.

3. Compare your results from Exercises 1 and 2. Does the model projection reproduce the data used to build the model? If it doesn't reproduce the 1984 data, what might explain the discrepancy?

HINTS: Be sure to save your files in ASCII text format. When running MARKOV, enter full file name (including extensions) and be sure input file is not also open in some other program.

EXERCISE 4
Model Validation

Model validation consists of testing a model against data that were *not* used to construct the model (Haefner, 1996). This is important as it is an *independent* test of the model.

1. Still using the 1972 data as your initial condition vector, use the program MARKOV to project the model to 1991 (i.e., 19 time steps).

2. Compare the predicted landscape composition to the actual composition tallied from the primary data table. This is a test to **validate** your model. Does the model projection match the 1991 data? If not, what might explain the discrepancy?

3. Continue to project the model into the future until it converges to a steady-state or until there is less than 10% of the landscape in old growth forest, whichever comes first. How long will the old-growth last, or when will it equilibrate?

NOTE: You can complete all of these tasks with a single projection of the model. Simply run the model for a very long time (say, 2000 years). If it converges in less time, simply delete the extraneous years from the model output file using a text editor (the output is only interesting while the landscape is changing).

CONCLUSIONS

This concludes the development, verification, and validation of a simple Markov model of landscape change. In some applications, such a simple model is sufficient (e.g., Johnson and Sharpe, 1976; Johnson, 1977; Hall et al., 1991). But in many cases, this simple model serves as a point of departure for more complicated models (e.g., Turner, 1987; Baker, 1989; Acevedo et al., 1995, 1996; Wear and Bolstad, 1998; Wu and Webster, 1998; Hong and Mladenoff, 1999a,b; Mladenoff and Baker, 1999, Urban et al., 1999). In particular, some consideration of the assumptions and limitations of Markov models can be a useful aid in interpreting the behavior and predictions of other models. These considerations are presented next.

Further Considerations in Modeling Landscape Change

The simple Markov model serves as a useful point of departure for more complicated issues in landscape modeling. Several of these are especially relevant to landscape dynamics.

Stochasticity

Models such as the one you have created are often used to project a map into the future by changing each cell in the map according to the transition probabilities. Because each cell can only change into one other state (a cell can't change fractionally), the state changes must be done probabilistically. This can create some new problems; the new map is only one of many possible stochastic realizations of a new map (because the map was created probabilistically). Thus, any comparison of model results to a real map would have to be based on a number of simulated replicate maps.

Importance of History

Another issue concerns the assumption that to predict the future state of a system one need only know its current state. In cases in which this is true, the process is truly first-order, also known as a Markov chain. In reality, there may be cases in which information about additional prior states is needed. These cases would lead to higher-order Markov models (e.g., in a second-order model, one would need to know the state of the system at time t and $t - 1$ to predict its state at time $t + 1$). Systems with an even longer "memory" would require still higher-order models. When the memory is very long, it might become more convenient to envision (and model) the dynamics in terms of a "time since" variable, such as "time since abandonment" or "time since disturbance," instead of keeping track of a large number of previous states. Chapter 12, Alternative Stable States, uses this approach, using a "time since fire" variable in a model of fire dynamics.

Stationarity

Because we have three maps, two first-order Markov models could be derived from our study landscape. We could derive transition matrices from 1972 to 1984, and another from 1984 to 1991. There is a formal test for stationarity of these matrices (Usher, 1992); nonstationary transition probabilities would vary among time periods. Nonstationary transition matrices would suggest that the forces (or rules) governing landscape change were changing over time. Certainly the drivers of landscape change might vary through time in regions with historical variation in socioeconomic drivers, for example, a condition that has existed over most of the United States over the past several decades.

In the case of nonstationary transition rules, two alternatives are possible. In the discrete case, separate transition matrices can be computed for each time period of interest. For example, given a sequence of airphotos taken every 10 years for 50 years, one could derive four separate transition matrices. Each matrix would be used to project from one time period to the next. Alternatively, the transitions could be specified explicitly as functions of time, so that the rules governing landscape change would vary with time. This approach would generate "smoother" dynamics, but would require some sort of curve-fitting for the time functions (as well as the data to support that curve-fitting!).

Spatial Dependencies

A fourth complication arises if some of the transitions appear to have spatial dependencies. For example, certain kinds of transitions might tend to occur in certain topographic settings or in certain spatial configurations as defined by a cell's immediate neighbors. These complications drive the modeling approach away from a simple Markov framework toward models in which the transition probabilities depend not only on the current state of the system, but also on some other stated conditions. That is, the transition matrix contains **conditional probabilities** such as "if the cell is type i and meets condition k, then its probability of becoming type j is p_{ijlk}." The condition might relate to site conditions (e.g., soil) or neighborhood effects (e.g., contagious disturbance). It is a relatively straightforward procedure to tally transition matrices as conditional probabilities. One simply constructs a multilayered tally matrix analogous to Table 4.4 that incorporates all of the special conditions of interest. Clearly, this can become extremely data-hungry. Again, similar conditional probabilities are examined in Chapter 12.

In any of these more complicated models of landscape change, the ability to solve the model analytically is rapidly lost; thus, complex models must be "solved" by iteration to steady-state (if such a state exists). Such a trade-off between simplicity and realism is common to all modeling efforts.

WRITE-UP

Your lab write-up should include the following sections:

1. The **Introduction** should state the motives of the exercise and provide some context. For example, why would we want to model landscape change?

2. The **Background** section should focus on the conceptual basis and assumptions of a first-order Markov chain as a model of landscape change. Address the following issues:
   (a) Try to present a Markov model, in a narrative sense, as clearly and concisely as possible. How does it work?
   (b) What assumptions do we make, implictly or explicitly, in using such a simple model?
   (c) Given the assumptions and the simplicity of such models, why use them at all? That is, what is the value of simple models in assessing landscape change?

3. The **Methods** should reiterate, concisely, the steps you followed to generate the model. This section could culminate in a presentation of the model as a transition matrix and as a graph, along with a table with the state vectors.

4. **Results** should consist of:
   (a) The projection of your landscape from 1972 to 1991 and your comparison of the model projection to the actual data for 1984 and 1991. You should be able to include all of this in one figure. You need *not* include the actual model output in tabular form.
   (b) Include your hand calculations for the one-year projection.

5. The **Discussion** should address the following questions:
   (a) How well does the model projection match the actual 1991 data? If it doesn't match, what possible reasons might you suggest for the discrepancy?

(b) How would you address these discrepancies (what changes to the model or what additional data would you need)?

(c) Would you expect the landscape to ever reach a steady-state? Of what interpretative value is the model solution (i.e., the steady-state composition of the landscape)?

(d) What would it take to maintain 20% of the landscape in old growth? That is, which transition rates would have to change, and how much?

ACKNOWLEDGMENTS

The classified imagery used in this lab was developed under funding provided by the USDA Forest Service New Perspectives Program, the National Sciences Foundation through the H. J. Andrews Long-Term Ecological Research (LTER) site (Grant 90-11663) and the Terrestrial Ecology Program of the National Aeronautics and Space Administration (NAGW-3745).

BIBLIOGRAPHY

Note. An asterisk preceding the entry indicates that it is a suggested reading.

ACEVEDO, M., D. L. URBAN, AND M. ABLAN. 1995. Transition and gap models of forest dynamics. *Ecological Applications* 5:1040–1055.

ACEVEDO, M. F., D. L. URBAN, AND H. H. SHUGART. 1996. Models of forest dynamics based on roles of tree species. *Ecological Modelling* 87:267–284.

*BAKER, W. L. 1989. A review of models of landscape change. *Landscape Ecology* 2:111–133. A general review and good introduction to a wider variety of models.

COHEN, W. B., M. FIORELLA, J. GRAY, E. HELMER, AND K. ANDERSON. 1998. An efficient and accurate method for mapping forest clearcuts in the Pacific Northwest using Landsat imagery. *Photogrammetric Engineering and Remote Sensing* 64:293–300.

COHEN, W. B., T. A. SPIES, AND M. FIORELLA. 1995. Estimating the age and structure of forests in a multiownership landscape of western Oregon, USA. *International Journal of Remote Sensing* 16:721–746.

FEMAT. 1993. Forest Ecosystem Management: An Ecological Economic, and Social Assessment, Report of Forest Ecosystem Management Assessment Team, July 1993, Portland, Oregon.

*HAEFNER, J. W. 1996. *Modeling Biological Systems: Principles and Applications.* Chapman and Hall, New York. The introductory chapter and Chapter 8 on model validation are great background reading.

HALL, F. G., D. B. BOTKIN, D. E. STREBEL, K. D. WOODS, AND S. J. GOETZ. 1991. Large-scale patterns of forest succession as determined by remote sensing. *Ecology* 72:628–640.

HANSEN, A. J., T. A SPIES, F. J. SWANSON, AND J. L. OHMANN, 1991. Conserving biodiversity in managed forests. *BioScience* 41:382–392.

HARRIS, L. D. 1984. *The Fragmented Forest.* University of Chicago Press, Chicago.

HE, S. H., AND D. J. MLADENOFF. 1999a. Spatially explicit and stochastic simulation of forest-landscape fire disturbance and succession. *Ecology* 80:81–99.

HE, S. H., AND D. J. MLADENOFF. 1999b. The effects of seed dispersal on the simulation of long-term forest landscape change. *Ecosystems* 2:308–319.

JOHNSON, W. C. 1977. A mathematical model of forest succession and land use for the North Carolina Piedmont. *Bulletin Torrey Botanical Club* 104:334–346.

JOHNSON, W. C., AND D. M. SHARPE. 1976. An analysis of forest dynamics in the northern Georgia Piedmont. *Forest Science* 22:307–322.

MARCOT, B. G., AND J. W. THOMAS. 1997. Of spotted owls, old growth, and new policies: A history since the Interagency Scientific Commmittee report. USDA Forest Service General Technical Report PNW-GTR- 408. Pacific Northwest Forest and Range Experiment Station, Portland, Oregon.

*MLADENOFF, D., AND W. L. BAKER, EDS. 1999. *Spatial Modeling of Forest Landscape Change: Approaches and Applications.* Cambridge University Press, Cambridge, UK. This book includes a good sampling of more recent approaches to modeling landscape dynamics.

ROBBINS, W. G. 1988. *Hard Times in Paradise, Coos Bay, Oregon, 1850–1986.* University of Washington Press, Seattle.

RUGGIERO, L. F., K. B. AUBRY, A. B. CAREY, AND M. H. HUFF (technical coordinators).1991. Wildlife and vegetation of unmanaged Douglas-fir forests. USDA Forest Service General Technical Report PNW-GTR-285. Pacific Northwest Forest and Range Experiment Station, Portland, Oregon.

SKLAR, F. H., and R. COSTANZA. 1991. The development of dynamic spatial models for landscape ecology: A review. In M. G. Turner and R. H. Gardner, eds. *Quantitative Methods in Landscape Ecology.* Springer-Verlag, New York, pp. 239–288.

SPIES, T. A., W. J. RIPPLE, AND G. A. BRADSHAW. 1994. Dynamics and pattern of a managed coniferous forest landscape in Oregon. *Ecological Applications* 4:555–568.

TURNER, M. G. 1987. Spatial simulation of landscape changes in Georgia: A comparison of 3 transition models. *Landscape Ecology* 1:29–36.

URBAN, D. L., M. F. ACEVEDO, AND S. L. GARMAN. 1999. Scaling fine-scale processes to large-scale patterns using models derived from models: Meta-models. In D. Mladenoff and W. Baker, eds. *Spatial Modeling of Forest Landscape Change: Approaches and Applications.* Cambridge University Press, Cambridge, UK, pp. 70–98.

USDA FOREST SERVICE AND USDI BUREAU OF LAND MANAGEMENT. 1994. Record of decision for amendments for Forest Service and Bureau of Land Management planning documents within the range of the northern spotted owl. USDA Forest Service and USDI Bureau of Land Management, Portland, Oregon.

*USHER, M. B. 1992. Statistical models of succession. In D. C. Glenn-Lewin, R. K. Peet, and T. T. Veblen, eds. *Plant Succession: Theory and Prediction.* Chapman and Hall, London, pp. 215–248. This tutorial on Markov processes and statistical models is an excellent introduction to this type of model and includes clearly worked examples as well as formal treatment of more formal analysis of Markov models.

WALLIN, D. O., M. E. HARMON, W. B. COHEN, M. FIORELLA, AND W. K. FERRELL. 1996a. Use of remote sensing to model land use effects on carbon flux in forests of the Pacific Northwest, USA. In H. L. Gholz, K. Nakane, and H. Shimoda, eds. *The Use of Remote Sensing in the Modeling of Forest Productivity at Scales from the Stand to the Globe.* Kluwer Academic Publishers, Dordrecht, the Netherlands, pp. 219–237.

WALLIN, D. O., F. J. SWANSON, B. MARKS, J. KERTIS, AND J. CISSEL. 1996b. Comparison of managed and pre-settlement landscape dynamics in forests of the Pacific Northwest, USA. *Forest Ecology and Management* 85:291–310.

WEAR, D. N., AND P. BOLSTAD. 1998. Land-use changes in southern Appalachian landscapes: Spatial analysis and forecast evaluation. *Ecosystems* 1:575–594.

WEINSTEIN, D. A., AND H. H. SHUGART. 1983. Ecological modeling of landscape dynamics. In H. A. Mooney and M. Godron, eds. *Disturbance and Ecosystems.* Springer-Verlag, New York, pp. 29–45.

WU, F., AND C. J. WEBSTER. 1998. Simulation of land development through the integration of cellular automata and multicriteria evaluation. *Environment and Planning B: Planning and Design* 25:103–126.

CHAPTER 5

SIMULATING CHANGES IN LANDSCAPE PATTERN

ERIC J. GUSTAFSON

OBJECTIVES

Landscapes are characterized by their structure (the spatial arrangement of landscape elements), their ecological function (how ecological processes operate within that structure), and the dynamics of change (disturbance and recovery). Understanding and predicting the dynamics of landscape change is an especially challenging aspect of landscape ecology. Landscape change is difficult to study because controlled experiments at landscape scales often are not feasible for political, economic, social, and logistical reasons. Opportunistic studies of change (e.g., after a large fire) are often confounded by uncontrolled factors. For these reasons, changes in landscape pattern are often studied using simulation models. This lab will

1. introduce simulation modeling as an important tool of landscape ecology;
2. show the utility of simulation models for examining landscape change at spatial and temporal scales that are not easily addressed using field methods;
3. illustrate an applied use of simulation modeling in landscape ecology— examining changes in landscape pattern caused by timber management;
4. discuss the assumptions and limitations of simulation models; and
5. show how models can be used to answer questions about landscape pattern and landscape change.

This exercise focuses on landscape change produced by forest management, using a timber harvest simulation model. The model you will use is a simplified version of HARVEST (Gustafson and Rasmussen, 2002), which has been shown to generate patterns similar to those produced by timber management (Gustafson and Crow, 1999). The model allows you to change the size of timber harvest openings, the total area harvested, and the spatial distribution of harvested areas (i.e., whether harvests will be clumped or dispersed). You will determine how different harvest regimes influence the amount of forest interior, the amount of forest edge, and the mean patch size of forests.

INTRODUCTION

Simulation Modeling

Science is a process of ruling out ideas that are *not* true, always leaving some uncertainty about the ideas we think *are* true. The things we accept as scientifically true are actually a collection of conceptual models of how we *believe* the world works. Formalizing a conceptual model using mathematical relationships, such as in a simulation model, can help us to make projections based on the relationships, assumptions, and initial conditions that we assume are reasonable for the model. Simulation models are used for a variety of reasons: to predict (e.g., tree-growth models), to improve our understanding of new theoretical models (e.g., metapopulation theory), and to illuminate how we might manage an ecological system (e.g., by timber harvest) to produce desired conditions. Simulation models also allow control of effects that are difficult to control in empirical experiments.

Spatial simulation models specifically include the spatial arrangement of key elements of the system being studied. Simulation modeling is especially suited to answer questions about the spatial implications of interacting processes, especially when manipulative experiments of many factorial combinations are not feasible. While some **stochastic** spatial models (i.e., based on a random process) are not useful in predicting the specific location of individual events, they can be used to generate overall patterns throughout a landscape, given a set of assumptions. These simulated patterns may be statistically indistinguishable from those that would be produced in the real world. However, if comparison of model results and empirical data are significantly different, we can conclude that our model does not adequately simulate reality.

Spatial modeling also allows identification of the parameters to which spatial pattern is most sensitive, focusing hypothesis testing and empirical model development. When using a simulation model, it is critical to understand the sensitivity of model results to changes in the input parameters (Haefner, 1996). Large changes in some parameters may have little effect on the model results, while small variation in other parameters may induce large effects on the results. Other parameters may have little effect, or the sensitivity of the model results may be related to the magnitude of the parameter value (a nonlinear relationship). An understanding of these properties of the model is gained by systematically varying input parameters, a process known as **sensitivity analysis.**

In this exercise you will conduct a limited sensitivity analysis of the HARVEST Lite model, which will also enable you to examine the relationship between a disturbance process (timber management) and landscape pattern. The model is a simplified version of HARVEST (Gustafson and Rasmussen, 2002). HARVEST was designed to simulate even-aged timber harvest techniques that generate a new stand of trees that are all the same age (e.g., clear-cutting, shelterwood, seed-tree techniques). The model was built based on knowledge of the past dispersion of timber harvests on the Hoosier National Forest (Indiana, USA) and has been shown to generate patterns similar to those produced by timber management (Gustafson and Crow, 1999). For the purposes of this exercise, the model was simplified to minimize the input data required and to allow you to experiment with the most interesting and important parameters without causing confusion by too much complexity. The model allows you to change the size of timber harvest openings, the total area harvested, and the spatial distribution of harvested areas (whether harvests will be clumped or dispersed).

Change in Spatial Pattern

Disturbance usually produces patches (i.e., distinct areas with habitat conditions that are different from surrounding areas). The patchiness of a landscape mosaic is the result of the interaction of past disturbance and the heterogeneity of the abiotic environment. The patch structure of landscapes is thought to have a significant effect on ecological communities (Turner, 1989). Consequently, monitoring change in patch-based measures of spatial pattern is an important way to assess landscape change.

An important consequence of intense disturbance (including even-age timber management techniques) in forested ecosystems is the creation of **edge habitat,** which is related to a reduction in **forest interior habitat.** A number of species appear to be sensitive to the presence of edge habitat (forest that is in proximity to a forest edge). It is not entirely understood why this sensitivity exists. It is likely related to the reduction in forest habitat found within circular home ranges located near open areas (King et al., 1997), although increased predation or brood parasitism rates in edge habitats have been observed (Brittingham and Temple, 1983; Andren and Anglestam, 1988). Conversely, some species prefer edge habitat, and their numbers respond positively to its creation (Litvaitis, 1993).

It is also not clear how far edge effects reach into forest habitat. Effects on vegetation (related to light and microclimate) may extend only a few tens of meters into the forest (Chen et al., 1992). For some forest interior birds, the effect may extend 100 to 500 meters into the forest (Andren and Anglestam, 1988; DellaSalla and Rabe, 1987; Van Horn et al., 1995), although the strongest evidence suggests the effect probably extends only about 50 meters

(Paton, 1994). Likewise, it is not known how far from an edge the habitat will still be suitable for edge species. Because of this uncertainty, it is useful to analyze the amount of edge and interior habitat using a range of edge buffer widths. The amount of interior present is quite sensitive to the width of the edge buffer under certain patterns of forest openings, as you will discover.

Change in spatial pattern is also related to the rate of recovery from disturbance. When recovery is quick, disturbance effects are more transient. Timber harvest openings are generally ephemeral—succession occurs and forests regrow. However, the rate of recovery may vary widely depending on a number of factors, most notably climate (precipitation and temperature) and soil conditions. For this reason, the persistence of disturbance effects may vary markedly among different parts of the world.

Understanding how spatial pattern in forests is affected by different harvesting strategies is facilitated by using spatial simulation models. The conceptual basis for simulation of harvest patterns at landscape scales can be traced back at least to the coarse-grid cutting model developed by Franklin and Forman (1987). Other similar pattern-generation models include LSPA (Li et al., 1993), Cascade (Wallin et al., 1994), Harvest (Gustafson and Crow, 1994, 1996), and the Dispatch model of Baker (1995). These models differ in the input data required and the sophistication of the scenarios they can simulate.

The HARVEST Lite Model

HARVEST Lite is a simple, yet powerful harvest simulator that allows manipulation of the most important determinants of spatial pattern in managed forests. Using maps derived from the Hoosier National Forest, HARVEST Lite allows the user to simulate 8 decades of forest harvest, alter the spatial pattern of harvests, and analyze the effects of the harvests on the amount of edge and interior habitat throughout the managed area. More specifically, HARVEST Lite allows the user to specify the definition of forest interior by specifying the width of the edge buffer used to calculate forest interior. The user also specifies the rate at which harvested areas function as openings in the forest interior by altering the time necessary for openings to recover to a closed canopy condition. To determine how the changes in these parameters affect spatial pattern, HARVEST Lite also analyzes the patch structure of the resulting forest age map. Patches are identified using an eight-neighbor rule, meaning that cells of the same age that share a common edge or corner are considered part of the same patch.

MODEL INPUT

Forest Age Map of Initial Conditions. Managed forests are typically divided into stands. A **stand** is an area that has a common history and is relatively homogeneous with respect to forest composition and age. The age of a stand usually reflects the time since a harvest or other disturbance such as fire or windthrow regenerated the forest. Two forest age (stand) maps derived from stand maps of the Hoosier National Forest are supplied for this exercise. One represents a managed landscape with stands ranging in age from <10 to 140 years old (**managed.gis**). The other (**undistbd.gis**) contains a map of a landscape with *no* young stands, suggesting a lack of disturbance. Because none of the stands in this map are too young to be harvested, there are initially no spatial constraints on harvest in this landscape. Nonforested areas appear black on the forest age maps; a lake occurs in the right center of the input maps provided, and a small agricultural area appears in the lower right corner. These maps represent an area of almost 4000 hectares, with a cell size of 30 meters (0.09 ha).

Mean Harvest Size (ha). This is the average size of harvests (in hectares) that HARVEST Lite will apply to the landscape. The model will generate harvest sizes from a distribution of sizes having this mean value and a standard deviation of 10% of the mean. In real-world management, values may range from <1.0 hectare to more than 300 hectares, depending on the ecosystem and the management goals. Mean harvest sizes range from 2.8 to 7.0 hectares in published management plans for the Hoosier National Forest. Harvests are only permitted in forest stands older than 8 decades.

Percent of Forested Area to Cut (per decade). This is the percent of the forested area in the input map that will be cut by the model *in each decade*. For example, if 10% of the forest is cut each decade, 80% of the forest will have been harvested by the end of the eight-decade simulation. The percent of forested area cut within timber production management areas ranges from 5.4 to 11.5% in the Hoosier National Forest.

Dispersion Method. Two spatial dispersion methods for placing harvests on the landscape are available: **dispersed**—all harvests are placed independently, or **clumped**—harvests are placed in clusters of nine openings. In both cases, harvests are only permitted in forest stands older than 8 decades in age.

Two additional parameters are specified for the analysis of forest interior and edge:

Edge Buffer Width (m). This is the maximum distance from a forest opening that edge conditions exist. Interior conditions are assumed to exist at distances greater than this value. Please note that HARVEST Lite must use a value that is a multiple of the cell width (in this case, 30 m) and will convert other values to the nearest multiple of the cell width. Values proposed to define habitat for forest interior birds range between 50 and 500 meters.

Opening Persistence Time (in decades). This is the time (in decades) that it takes for harvest openings to regrow to closed canopy conditions. Harvested cells younger than this value are considered openings, while cells older than this value are assumed to have a closed canopy. Canopies may become closed within 12 to 15 years of clearcutting in some ecosystems with productive soils and moist climate, but may require several decades in areas with poorer soils or drier climates.

MODEL OUTPUT

Each simulation represents eight decades of harvest activity. Model outputs take the form of maps and map analysis reports, including the following:

Forest Age Map. This map is displayed upon completion of the simulation and reflects the cells harvested during the simulation and the ages of unharvested cells. This map may be saved and used as input for other simulations.

The patches analysis produces a variety of outputs available in the log file. One of the more important results is explained next.

Mean Size of Patches (ha). This is the average size of patches, with *patches* defined as contiguous cells of the same forest age. Some of these patches will be the result of simulated harvests, but some will be remnants of the uncut forest. Consequently, the mean size of patches will not likely equal the mean harvest size you used to simulate harvest activity.

The interior analysis produces the following two results:

Area of Interior Habitat (ha). This is the area of forest interior conditions calculated based on the forest age map. Interior conditions are defined by the edge buffer width described earlier. The forest interior map produced by HAR-VEST Lite shows forest interior habitat in red. Note that a measure of the total boundary length between different-aged cells (km) (a measure of *linear* edge) is calculated as part of a patch analysis, and this is different from the *area* of edge habitat calculated as part of the interior analysis.

Area of Edge Habitat (ha). This is the area of forest edge conditions calculated based on the forest age map. Interior conditions are defined by the edge buffer width described earlier. The forest interior map produced by HARVEST Lite shows forest edge habitat in various colors other than red.

A number of simplifying assumptions were made in the development of HAR-VEST Lite to reduce input data requirements and to enable it to simulate harvest activity quickly over a relatively large area. Based on harvest activity in the Hoosier National Forest (Gustafson and Crow, 1996), it is assumed that harvest allocations are randomly distributed within timber management areas when accumulated over the course of a decade. However, HARVEST Lite operates within the constraint that harvests cannot be placed where the forest is younger than 80 years, and all simulations run for eight decades. HARVEST Lite uses stand age as a surrogate for merchantability and ignores stocking density and size class.

Several other simplifications have been made to reduce model complexity for this exercise. Although you may specify the mean size of harvests, the standard deviation around this mean has been fixed at 10% of the mean harvest size. When HARVEST Lite produces clumped distributions, the nucleus of each clump is randomly placed, and then eight other harvest units are placed randomly around the initial harvest. HARVEST Lite always leaves a one-cell buffer between harvests allocated in the same decade, and between harvests and any nonforested land uses. HARVEST Lite ignores specific forest types, assuming

that forest types are harvested in proportion to their availability. The proximity of roads and the feasibility of conducting logging operations are assumed
to be uniform across the land base.

Instructions for Using HARVEST Lite

Start HARVEST Lite by double-clicking on its icon (or HarvLite.exe). Online help
is available from the HARVEST Lite pull-down **Help** menu. *NOTE: Be sure to
copy the entire folder for this chapter to your computer as HARVEST Lite will
not run directly from the CD.* If you encounter problems, ensure that the
upper-level directory names (above the folder containing HarvestLite) do not
contain spaces or dashes, and that the full pathname is less that 100 characters.

1. Specify the base forest age map on which you will conduct the simulations by selecting **"Choose base map"** from the **Model** menu. This will
   load the map into memory and allow you to analyze the initial pattern,
   or alternatively, immediately begin a simulation.
2. The spatial pattern of patches and forest interior on a map may be analyzed at any time by making the appropriate choice from the **Analyze** menu.
   (a) **Patches analysis:** An analysis of patches will calculate the mean size
       of patches (defined by forest age). The results of this analysis will
       be printed to the screen and are written to a running log file that
       can be saved as a record of your analysis.
   (b) **Interior analysis:** When you analyze forest interior, you will be asked
       to provide an edge buffer width and an opening persistence time.
       HARVEST Lite will calculate the amount of interior and edge habitat
       based on these values and will display a map of forest interior and
       edge. You may conduct multiple analyses of interior on the same
       forest age map using various interior definition values. These analy-
       ses are also written to the running log.
3. You can save your results as follows:
   (a) **Log file:** The running log file will be displayed on screen, under
       "Progress & Results." The running log can be saved to a text file
       at any time using the **"Save log file"** option under the **Save** menu.
       The running log is cleared when you load a new base map (and when
       you save the log file), so if you wish to save any analyses, do so
       prior to loading a base map.
   (b) **Interior map:** The interior map may also be saved by selecting the
       "Save interior map" option. Map files are saved in ERDAS 7.4 for-
       mat and may be loaded into many common GIS systems or used as
       input maps for other HARVEST Lite simulations.
4. To conduct a simulation of harvesting, select **"Execute"** from the **Model**
   menu. Next, you can enter a random number seed, or simply enter 0 to
   have the computer pick a random number for you. A dialog will allow
   you to set the parameters that control how HARVEST Lite allocates harvests on the landscape, including mean harvest size, percent of forested

area to cut each decade, and the dispersion method for harvests (dispersed or clumped). When HARVEST Lite finishes the simulation, the updated forest age map will be displayed. This new age map can be saved by selecting **"Age map"** under the **Save** menu.

5. You may now analyze the pattern of this changed landscape using the analysis functions described in step 2. These analyses will be appended to those you may have conducted previously. You may also wish to save the new maps for later analysis, or to use as input for further simulations.

6. To conduct a new simulation using different parameters, reload a base map by selecting **"Choose base map"** under the **Model** menu. This will clear all prior maps, analyses, and parameter settings from memory.

7. To quit HARVEST Lite, choose **"Exit"** from the **Model** menu.

Forest Harvest Simulation Scenarios

Complete the simulations for all assigned exercises before creating graphs and answering discussion questions. This will ensure that you complete the simulations in the allotted time. Keep in mind that this is a stochastic simulation model (i.e., simulations are based on random number sequences). You will not get exactly the same results on successive runs, and your results will differ slightly from those of your classmates.

EXERCISE 1
Effects of Mean Harvest Size

Forest managers are being compelled (either by regulation or public opinion) to reduce the size of clearcuts and other timber harvest activities. For example, there is a 16-hectare limit on the size of clearcuts in most national forests. One might expect that smaller clear-cuts are less disruptive to forest habitats and will leave more of the forest free from the effects of disturbance. This exercise will examine the effect of changing mean harvest size on forest spatial structure.

1. From the Model menu, select **"Choose base map."** Use the input file **managed.gis,** found in the same directory as the HARVEST Lite program itself.

2. Use HARVEST Lite to simulate four forest management scenarios in which mean harvest sizes vary (using sizes of 1, 10, 20 and 30 ha). For the first run, select **"Execute"** from the Model menu. Enter a **Mean harvest size** of 1.0. Verify that the values for the other two simulation parameters (percent of forested area to cut and dispersion method) are held constant at: **Percent of forested area to cut = 3.0,** and **"Dispersed"** for the dispersion method. Click on the OK button to start the simulation.

3. When the simulation has completed, calculate the amount of forest interior by selecting **"Interior (after harvest)"** from the **Analyze** menu. Enter an **Edge buffer width** of 180 meters and an opening persistence time of 2 decades. Be sure to use these values for each simulation in this exercise. Also, conduct a patch analysis by selecting **"Patches (after harvest)"** from the **Analyze** menu.

4. Use Table 5.1 on the CD to record the mean harvest size, area of interior habitat, area of edge habitat, and mean patch size (across all age classes) for each run. These values can be found in the **Progress and Results** window after each analysis is completed. If you wish to save the log file after each simulation, be sure to do it prior to reloading the base map (select **"Save log file"** under the **Save** menu).

5. Repeat steps 1–4 for the other three harvest sizes (10, 20, and 30 ha).

6. Use the data recorded in Table 5.1 to plot area of interior, edge, and mean patch size against mean harvest size. You may make the plot on graph paper or use a spreadsheet program if one is available. Write a caption below each plot and indicate the values of the parameters that were held constant.

Question 1.1. Is there a threshold effect (i.e., a small range of values within which the effect changes markedly) of mean harvest size? If so, at approximately what mean harvest size does the threshold occur?

Question 1.2. If you were advising a forest manager who was under pressure to both minimize harvest size and maximize forest interior habitat, what would you recommend as a policy for mean harvest size?

EXERCISE 2
Effects of Percent of Forest Cut Each Decade

Timber production levels are declining on many publicly owned forests, primarily to enhance biodiversity and other noncommodity values of forests. This is primarily caused by a reduction in the percentage of the land where timber harvest is allowed. In this exercise we will examine the effect of changing the percent of forest cut each decade.

1. From the Model menu, select "Choose base map." Use the input file **managed.gis**.

2. Use HARVEST Lite to simulate four forest management scenarios in which the percent of forested area to cut varies, from 1, 3, 5, to 7% of the landscape each decade. Hold the other parameters constant for each of these runs. Use a mean harvest size of 5.0 hectares, and the "Dispersed" dispersion method.

3. Analyze the map to determine the amount of forest interior using an edge buffer width of 180 meters and an opening persistence time of

2 decades. Be sure to use these values for each simulation in this exercise. Also, conduct a patch analysis.

4. Record the area of interior habitat, area of edge habitat, and mean patch size (across all classes) in Table 5.2 on the CD.

5. Repeat steps 1–4 for each percent of forested area to cut (1, 3, 5, 7%).

6. Use the data in Table 5.2 to plot the area of interior habitat, the area of edge habitat, and the mean patch size (across all classes) against percent of forested area cut. Write a caption below each plot and indicate the values of the parameters that were held constant.

7. (Optional) If you have access to graphing software, you may wish to produce a 3-D surface plot combining the results of this and the previous exercise. Additional simulations will be necessary to complete the plot. Compare your results to those in Gustafson and Crow (1994).

Question 2.1. What is different about the shape of these plots compared to those generated for the effects of mean harvest size?

Question 2.2. Does there appear to be a threshold effect of percent of forested area cut each decade? If so, at approximately what percent does the threshold occur?

EXERCISE 3
Effects of Spatial Dispersion

Forest interior habitat is thought to be important for a number of species that are declining in abundance. Several researchers have suggested that clustering harvest activity will increase forest interior habitat. Clustering serves to aggregate the edge effects of harvest openings, leaving larger blocks of contiguous forest. This exercise will examine the effects of clustering of harvests on area of forest interior, area of forest edge, and mean patch size.

1. Select the input file **undistbd.gis.**

2. Choose a mean harvest size between 5 and 30 hectares, and percent of the forested area to cut between 1 and 7%. Run three replicate simulations (parameters unchanged) each for dispersed and clustered harvests.

3. Analyze the map to determine the amount of forest interior using an edge buffer distance of 180 meters and an opening persistence time of 2 decades. Also, conduct a patch analysis.

4. Record data in Table 5.3 on the CD.

5. Produce three bar charts showing the mean area of forest interior, the mean area of edge, and the mean patch size (across all classes) for clustered and dispersed harvest patterns. Include an error bar for

each histogram bar, the length of the error bar being twice the standard error of the mean. Calculate standard error (*se*) using:

$$se = \sqrt{\frac{\sum\limits_{i=1}^{n}(y_i - \bar{y})^2}{n-1}} \bigg/ \sqrt{n}$$

where *n* is the number of replicate simulations (in this case, three), y_i is the mean (of interior or edge or patch size) of the *i*th replicate, and $\bar{y}$ is the mean of the three means (the y_is). Write a caption below each plot and indicate the values of the parameters that were held constant.

Question 3.1. Would you say that clustering harvests significantly changes the area of forest interior habitat? Forest edge habitat? Mean patch size?

Question 3.2. How did you judge significance from these plots? (HINT: Look at the error bars.)

EXERCISE 4
Effects of Edge Buffer Width and Opening Persistence Time

There is some debate about how far into the forest the effects of edge are evident. The effects related to reduced nesting bird densities and increased nest predation may extend much farther into the forest than microclimate effects. Forests also recover from harvesting at different rates in different ecosystems. Forests on good soils in moist climates may recover more quickly than forests on poor sites or in relatively dry climates. This exercise will examine how the spatial pattern of forest interior depends on how interior habitat is defined.

1. Select the input file **managed.gis.**
2. Simulate harvests with a mean harvest size of 1.0 hectare and 4% of the forest area cut each decade. Use the "Dispersed" dispersion method. This will be the only simulation run for this exercise.
3. Analyze the map to determine the area of forest interior using no edge width buffer (0 m), a 120-meter edge buffer width, and a 300-meter edge buffer width. Assume that openings persist for two decades. Do not conduct a new simulation between calculation of interior for each edge buffer width.
4. Record the data requested in Table 5.4 on the CD.
5. Without running another simulation, repeat the three calculations in step 3 using an opening persistence time of three decades.
6. Produce a bar chart of the results.

Question 4.1. Does an increase in edge buffer width have a disproportionate effect on the area of forest interior habitat? Why or why not?

Question 4.2. What is the effect of increasing the opening persistence time?

DISCUSSION QUESTIONS

1. What are the assumptions made by the HARVEST Lite model? Under what scenarios might they be reasonable or unreasonable? How might you test these assumptions? How does knowledge of the assumptions influence interpretation of the results?
2. Is forest interior more sensitive to variation in harvest size or percent of forested area to cut? To which is area of edge habitat more sensitive? To which is patch size more sensitive?
3. Was there a parameter missing from the model with which you wanted to be able to experiment? What was it, and why were you interested in it?
4. How did this exercise change your thinking about the spatial aspects of timber harvesting? How might the results of your simulations be used to develop a research project?
5. Consider each characteristic of landscape change represented by the parameters that can be manipulated by HARVEST Lite (mean size of harvests, percent of forest cut, dispersion). How are these characteristics related to change in other types of landscapes?

BIBLIOGRAPHY

Note. An asterisk preceding the entry indicates that it is a suggested reading.

ANDREN, H., AND ANGELSTAM, P. 1988. Elevated predation rates as an edge effect in habitat islands: Experimental evidence. *Ecology* 69:544–547.

BAKER, W. L. 1995. Long-term response of disturbance landscapes to human intervention and global change. *Landscape Ecology* 10:143–159.

BRITTINGHAM, M. C., AND S. A. TEMPLE. 1983. Have cowbirds caused forest songbirds to decline? *BioScience* 33:31–35.

CHEN, J., J. F. FRANKLIN, AND T. A. SPIES. 1992. Vegetation responses to edge environments in old-growth Douglas-fir forests. *Ecological Applications* 2:387–396.

DELLASALLA, D. A., AND D. L. RABE. 1987. Response of least flycatchers *Empidonax minimus* to forest disturbances. *Biological Conservation* 41:291–299.

*FRANKLIN, J. F., AND R. T. T. FORMAN. 1987. Creating landscape patterns by forest cutting: Ecological consequences and principles. *Landscape Ecology* 1:5–18. This paper was among the first to use a simulation model to investigate the landscape pattern effects of timber harvesting.

*GUSTAFSON, E. J. 1996. Expanding the scale of forest management: Allocating timber harvests in time and space. *Forest Ecology and Management* 87:27–39. In this paper Harvest was used to simulate several clustered cutting strategies on a real landscape encompassing the entire Hoosier National Forest in Indiana. Results showed that the area harvested could be increased while also increasing the amount of forest interior when a clustered strategy was used.

GUSTAFSON, E. J. 1998. Quantifying landscape spatial pattern: What is the state of the art? *Ecosystems* 1:143–156.

GUSTAFSON, E. J., AND L.V. RASMUSSEN. 2002. Assessing the spatial implications of interactions among strategic forest management options using a Windows-based harvest simulator. *Computers and Electronics in Agriculture* 33:179–196.

GUSTAFSON, E. J., AND T. R. CROW. 1996. Simulating the effects of alternative forest management strategies on landscape structure. *Journal of Environmental Management* 46:77–94.

GUSTAFSON, E. J., AND T. R. CROW. 1999. HARVEST: linking timber harvesting strategies to landscape patterns. In D. J. Mladenoff and W. L. Baker, eds. *Spatial Modeling of Forest Landscapes: Approaches and Applications*. Cambridge University Press, Cambridge, UK, pp. 309–332.

GUSTAFSON, E. J., AND T. R. CROW. 1994. Modeling the effects of forest harvesting on landscape structure and the spatial distribution of cowbird brood parasitism. *Landscape Ecology* 9:237–248.

*HAEFNER, J. W. 1996. *Modeling biological systems: Principles and applications*. Chapman and Hall, New York. This book provides an excellent introduction to simulation modeling.

KARPLUS, W. J. 1983. The spectrum of mathematical models. *Perspectives in Computing* 3:4–14.

KING, D. I., C. R. GRIFFIN, AND R. M. DEGRAAF. 1997. Effect of clearcut borders on distribution and abundance of forest birds in northern New Hampshire. *Wilson Bulletin* 109:239–245.

LEVIN, S. A. 1992. The problem of pattern and scale in ecology. *Ecology* 73:1943–1967.

*LI, H., J. F. FRANKLIN, F. J. SWANSON, AND T. A. SPIES. 1993. Developing alternative forest cutting patterns: A simulation approach. *Landscape Ecology* 8:63–75. The authors use a harvest simulation model to investigate novel strategies to spatially allocate timber harvests. For example, they evaluate a "progressive cutting" strategy in which harvest activity proceeds systematically across a landscape, as an alternative to traditional dispersed methods.

LITVAITIS, J. A. 1993. Response of early successional vertebrates to historic changes in land use. *Conservation Biology* 7:866–873.

PATON, P. W. 1994. The effect of edge on avian nest success: How strong is the evidence? *Conservation Biology* 8:17–26.

TURNER, M. G. 1989. Landscape ecology: The effect of pattern on process. *Annual Review of Ecology and Systematics* 20:171–197.

VAN HORN, M. A., R. M. GENTRY, AND J. FAABORG. 1995. Patterns of ovenbird (*Seiurus aurocapillus*) pairing success in Missouri forest tracts. *Auk* 112:98–106.

*WALLIN, D. O., F. J. SWANSON, AND B. MARKS. 1994. Landscape pattern response to changes in pattern generation rules: Land-use legacies in forestry. *Ecological Applications* 4:569–580. The authors used a harvest simulation model to show that landscape patterns produced by dispersed disturbances are difficult to erase, persisting in some form for a long time.

CREATING LANDSCAPE PATTERN

HAZEL R. DELCOURT

OBJECTIVES

Landscape pattern results from physical, biological, and cultural processes acting simultaneously over a broad geographic region. Geology, topography, soils, disturbance regimes, and land use all influence the configuration and spatial relationships of landscape elements as well as the individual distributions of species within the landscape mosaic. This lab is developed around a study area in the eastern Upper Peninsula of Michigan, which presents a heterogeneous landscape that largely reflects its complex glacial and post-glacial history.

This chapter consists of two separate lab exercises that both include (a) analyzing landscape pattern through analysis of mapped cover types and (b) determining underlying causes of landscape heterogeneity through examining distributions of individual tree species. The specific objectives for these labs are to:

1. quantify mapped landscape patterns and examining their relationship to land use history. You will accomplish this objective by comparing two thematic maps portraying pre-European settlement vegetation (mapped circa A.D. 1850 by U.S. General Land Office surveyors) with late 20th century land use (mapped circa A.D. 1980). For this objective, you will determine how the vegetation patchwork has changed from pre-European settlement times to the present by comparing kinds, numbers, sizes, and configurations of cover types.

2. determine underlying causes of landscape heterogeneity. To accomplish this second objective, you will compare thematic maps portraying the distributions of soils (characterized by pH, texture, organic content, and presence/absence of a hardpan or "ortstein" layer) and of disturbance regime (wildfire and Indian-set fire) with maps depicting the pre-European settlement distributions of individual species of trees.

The two sections of this lab are *independent* alternatives, not *sequential* options. For the basic option (Part I), you may calculate and plot all necessary statistics directly by hand. The advanced option (Part 2) uses FRAGSTATS metric calculation software for quantifying landscape patterns. The advanced option assumes a strong facility with FRAGSTATS and/or the completion of Chapter 7, Understanding Landscape Metrics I (a strongly recommended prerequisite). The materials provided include (a) thematic maps (Figures 6.2a–6.10a), (b) raster maps number-coded according to thematic maps (Figures 6.2b–6.10b), (c) and one blank raster map to create transparencies (Figure 6.11). Much larger versions of all the maps listed in (a), (b), and (c) have been provided on the CD. It is recommended that you print these for use in hand calculations. Finally, digital files, including the FRAGSTATS program and raster maps in FRAGSTATS-compatible ASCII text format (*.fra) are also on the CD. For both labs, you will also need a variety of different colored pencils and highlighter pens, graph paper, a calculator, and access to a photocopy machine.

Note regarding Fragstats 2.0 and Windows XP
The version of Fragstats included on the CD (FRAGSTATS 2.0) is not compatible with Windows XP. Using Fragstats on Windows XP requires downloading the latest shareware version of Fragstats 3.0 from the web at
  http://www.umass.edu/landeco/research/fragstats/fragstats.html

INTRODUCTION

The study area (Figure 6.1) is centered on the present-day village of Naubinway, Michigan, which was originally the site of a historic Native American fishing village located along the most northern part of Lake Michigan. The study area covers a total land area of 34,851 hectares in Garfield Township, western Mackinac County, Michigan. A large inland lake, Lake Millecoquins, is located approximately in the center of the study area.

Soils (Figure 6.2)

The modern physical landscape is a complex mosaic of geomorphic features formed by geological processes ranging from decades to thousands of years. Limestone bedrock underlies the region and is exposed locally west of Lake Millecoquins. Landforms such as hummocky moraines composed of glacial till—a mixture of cobbles, gravels, sands, and clays—were deposited by moving glacial ice over much of the study area between 18,000 and 11,000 years ago. Sand and gravel were laid down in deltas formed by rivers that carried glacial outwash from the melting margin of the continental glacier and de-

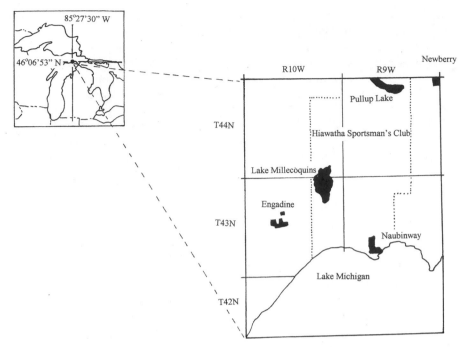

FIGURE 6.1
Location map of the study area, showing place names mentioned in the lab exercise.
The dashed line delineates the boundaries of the Hiawatha Sportsman's Club.
North–south and east–west lines indicate boundaries between 36-square-mile town-
ships (blocks six miles on a side); township and range coordinates are indicated by a
T or an *R* followed by number designations that relate to principal meridians.

posited these coarsely textured sands at the edge of a great freshwater sea, the
ancestor of present-day Lake Michigan. After 11,000 years ago, a series of
shorelines formed at elevations controlled by the changing water level of Lake
Michigan, driven by climate change, rising land, and cutting of new drainage
outlets to the sea (Petty et al., 1996). Since 5400 years ago, a broad lake plain
has formed south of Lake Millecoquins as Lake Michigan has receded south-
ward (Delcourt et al., 1996).

Soil cover types generally reflect underlying parent material (Whitney et al.,
1995). Xeric (well-drained), acidic (pH < 7), sandy soils (soil types 1 and 2
on map) occur on deposits created by the action of flowing river water at the
former margin of glacial ice. Sandy soils more than 4000 years old (soil type
2) are highly leached, with a dense, iron-cemented hardpan or "ortstein" layer
that impedes downward percolation of rainwater and results in a greater mois-
ture-holding capacity than sandy soils without ortstein development (soil type
1). Mesic (moderately drained) and hydric (poorly drained), calcareous (cal-
cium-rich), alkaline (pH > 7) soils (map units 3 and 5) developed over limestone bedrock or on lake clay deposits that incorporate calcium carbonate
leached from limestone. Mesic, acidic, sandy loam (map unit 4) formed over
glacial till. Hydric, acidic sand and mucky peat (map unit 6) are characteris-

a b

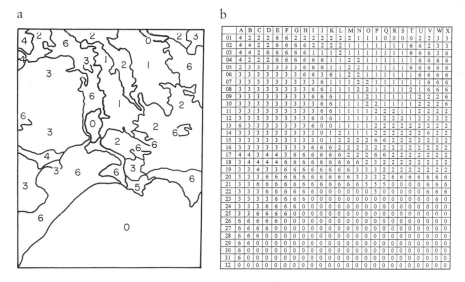

FIGURE 6.2
(a) Soil type map, based on Whitney et al. (1995), showing the distribution of soils grouped by similarities in pH, texture, organic content, and presence/absence of "ortstein" (a buried layer of cemented hardpan composed of aluminum-magnesium sesquioxides). Soil cover types are designated as follows:
0 = lake;
1 = xeric, acidic sand without ortstein;
2 = xeric, acidic sand with ortstein;
3 = mesic, calcareous, alkaline clayey to sandy loam;
4 = mesic, acidic, sandy loam;
5 = hydric, calcareous, alkaline gravelly loam;
6 = hydric, acidic sand and mucky peat.
(b) Raster grid derived from soils type map 6.2a with a resolution of 0.5 × 0.5 mile (0.25 mile^2). Cover types are designated using a 50% cover rule (the cover type is >50% of the area of the grid cell or it is the largest cover type within the grid cell). The FRAGSTATS-compatible (ASCII) file is designated **soils.fra** on the CD.

tic of Lake Michigan shoreline deposits that formed a poorly drained coastal lake plain in the past 5400 years, as well as of wetlands that have expanded farther inland over level terrain during the cool, moist climatic conditions of the past 3200 years (Delcourt and Delcourt, 1996).

Pre-European Settlement Vegetation (Figure 6.3)

The study area is located in a vegetation region that is transitional between temperate deciduous forest to the south and boreal coniferous forest to the north. The natural vegetation is sometimes referred to as the mixed conifer-northern hardwood forest or the hemlock-white pine-northern hardwood forest region (Delcourt and Delcourt, 1996).

U.S. General Land Office Survey (GLOS) records from the mid-19th century provide a description of vegetation that existed before settlement and log-a

b

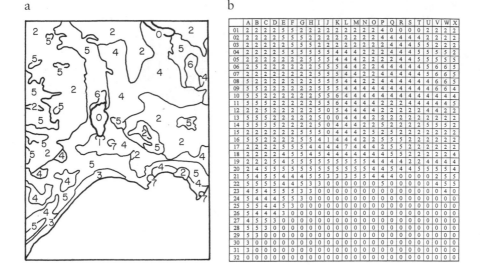

	A	B	C	D	E	F	G	H	I	J	K	L	M	N	O	P	Q	R	S	T	U	V	W	X
01	2	2	2	2	5	5	2	2	2	2	2	2	2	2	4	0	0	0	0	2	2	2	2	2
02	2	2	2	2	5	5	5	2	2	2	2	2	2	2	2	4	4	4	4	2	2	2	2	2
03	2	2	2	2	2	5	5	5	2	2	2	2	2	2	2	4	4	4	5	5	2	2	2	2
04	2	2	2	2	5	5	5	5	5	5	4	4	4	2	2	2	4	4	4	5	5	5	5	2
05	2	2	2	2	2	2	2	5	5	5	4	4	4	2	2	2	2	4	4	5	5	5	5	5
06	2	5	2	2	2	2	2	2	5	5	5	4	4	2	2	4	4	4	4	5	6	6	6	5
07	2	2	2	2	2	2	2	2	5	5	5	4	4	2	2	4	4	4	4	5	6	6	6	5
08	5	2	2	2	2	2	2	2	5	5	5	4	4	2	2	4	4	4	4	4	6	6	6	5
09	5	5	2	2	2	2	2	2	5	5	5	4	4	4	4	4	4	4	4	4	6	6	6	4
10	5	5	5	2	2	2	2	2	2	5	5	6	4	4	4	4	4	4	4	4	4	4	4	4
11	5	5	5	2	2	2	2	2	2	5	5	6	4	4	4	4	2	2	2	4	4	4	4	5
12	2	2	5	2	2	2	2	2	2	5	0	5	4	4	4	2	2	2	2	2	4	4	2	2
13	5	5	5	5	2	2	2	2	2	5	0	0	4	4	4	2	2	2	2	2	2	2	2	2
14	5	5	5	5	5	2	2	2	2	5	0	4	4	4	2	2	5	2	2	2	2	5	5	5
15	2	2	2	2	2	2	2	2	5	5	5	0	4	4	4	2	5	2	5	2	2	2	2	2
16	5	5	5	2	2	2	2	5	5	4	1	1	4	4	4	2	2	5	5	5	2	2	2	2
17	2	2	2	2	5	5	5	4	4	4	4	7	4	4	2	5	2	5	2	2	2	2	2	2
18	2	2	2	2	4	5	5	5	4	5	4	4	4	4	4	4	4	5	5	2	2	2	2	4
19	2	2	2	5	4	5	5	5	5	5	5	5	5	5	4	4	4	4	2	2	4	4	4	4
20	2	4	5	5	5	5	5	5	5	5	5	5	5	5	4	4	4	5	5	4	4	4	4	4
21	5	4	5	5	4	4	4	5	5	3	3	3	5	5	4	4	4	0	0	0	0	2	5	5
22	5	5	5	5	4	4	5	3	3	0	0	0	0	0	0	5	0	0	0	0	0	4	5	5
23	4	5	4	5	5	5	3	3	0	0	0	0	0	0	0	0	0	0	0	0	0	0	4	0
24	5	4	4	4	5	5	3	0	0	0	0	0	0	0	0	0	0	0	0	0	0	0	0	0
25	5	5	5	4	4	5	3	0	0	0	0	0	0	0	0	0	0	0	0	0	0	0	0	0
26	5	4	4	4	3	0	0	0	0	0	0	0	0	0	0	0	0	0	0	0	0	0	0	0
27	4	5	5	3	0	0	0	0	0	0	0	0	0	0	0	0	0	0	0	0	0	0	0	0
28	5	5	5	3	0	0	0	0	0	0	0	0	0	0	0	0	0	0	0	0	0	0	0	0
29	5	3	0	0	0	0	0	0	0	0	0	0	0	0	0	0	0	0	0	0	0	0	0	0
30	3	0	0	0	0	0	0	0	0	0	0	0	0	0	0	0	0	0	0	0	0	0	0	0
31	3	0	0	0	0	0	0	0	0	0	0	0	0	0	0	0	0	0	0	0	0	0	0	0
32	0	0	0	0	0	0	0	0	0	0	0	0	0	0	0	0	0	0	0	0	0	0	0	0

FIGURE 6.3
(a) Pre-European settlement vegetation map derived from the field notes and plat maps of U.S. General Land Office Surveys of 1840–1849 (Delcourt and Delcourt, 1996). Vegetation cover types are designated as follows:
0 = lake;
1 = cultural feature (Native American village);
2 = upland deciduous forest;
3 = upland evergreen forest;
4 = mixed deciduous/evergreen forest;
5 = forested wetland;
6 = nonforested wetland;
7 = Native American meadow.
(b) Raster grid derived from Figure 6.3a using a resolution of 0.5 × 0.5 mile (0.25 mile^2). Cover types are designated as in 6.3a, using 50% cover rule. The FRAGSTATS-compatible (ASCII) file is designated **pre.fra** on the CD.

ging by European Americans. Table 6.1 shows statistics for the area occupied by each vegetation type. GLOS plat maps have a minimum spatial resolution of 4 hectares and identify boundaries between upland vegetation and wetlands, positions of streams and lakes, and locations of witness and bearing trees used to mark section corners within the 36-square-mile grid of each surveyed township (Figure 6.1). For a discussion of methods employed in the GLOS and an evaluation of the grain of analysis appropriate for measuring landscape heterogeneity from GLOS data, see Delcourt and Delcourt (1996).

Xeric uplands generally supported evergreen or mixed deciduous/evergreen forest (cover types 3 and 4). Some of these forests were mixtures of old-growth white pine (*Pinus strobus*), paper birch (*Betula papyrifera*), and aspen (*Populus* spp.) on extensive sand plains (cover type 4). Other pine stands were

TABLE 6.1
AREA STATISTICS FOR PRE-EUROPEAN SETTLEMENT VEGETATION,
WESTERN MACKINAC COUNTY, MICHIGAN (FROM DELCOURT AND
DELCOURT, 1996). NUMBERS FOR COVER TYPES ARE FROM FIGURE 6.3.

Vegetation Type	Cover Type	Area (ha)	Area (%)
Cedar-spruce-tamarack-alder swamp	5	8871	25.5
White pine-paper birch-aspen forest	4	6669	19.1
Hemlock-hardwood forest	2	5532	15.9
Sugar maple-fir-beech-yellow birch forest	2	5313	15.2
Sugar maple-elm-ironwood-lynn (basswood) forest	2	2918	8.4
Jack pine-aspen forest	4	2545	7.3
Red pine forest	3	1263	3.6
Spruce-tamarack swamp	5	1105	3.2
Open marsh, meadow, and bog	6 & 7	635	1.8

dominated by jack pine (*Pinus banksiana*) with aspen on recently burned sites (cover type 4) or by red pine (*Pinus resinosa*) on more fire-protected sites along lakeshores (cover type 3).

Mesic, upland deciduous forest (cover type 2) was characteristic of pH-neutral and alkaline calcareous soils. The deciduous forest was dominated by sugar maple (*Acer saccharum*), American beech (*Fagus grandifolia*), elm (*Ulmus americana*), ironwood (*Ostrya virginiana*), basswood (*Tilia americana*), and yellow birch (*Betula alleghaniensis*). In some areas, deciduous trees grew in mixtures with evergreen conifers such as hemlock (*Tsuga canadensis*) and fir (*Abies balsamea*).

Two distinctively different kinds of plant communities were characteristic of wetlands (cover types 5 and 6). Forested wetlands (cover type 5), dominated by northern white cedar (*Thuja occidentalis*), black spruce (*Picea mariana*), tamarack (*Larix laricina*), and alder (*Alnus rugosa*), were extensive in drainageways of large inland lakes and on ancient shorelines within the Lake Michigan lake plain. Nonforested wetlands (cover type 6) were open marshes and bogs, probably dominated by herbaceous sedges (Cyperaceae) and cat-tails (*Typha* spp.) on pH-neutral to alkaline soils and by peat moss (*Sphagnum*) and thickets of cranberry and blueberry heaths (*Vaccinium* spp.) over acidic substrates.

Native Americans used fire to maintain open, grassy meadows (cover type 7) along trails leading to inland "sugar camps" (where they harvested maple syrup) and near seasonal fishing camps along Lake Michigan (Delcourt and Delcourt, 1996; Silbernagel et al., 1997).

Pre-European Settlement Tree Distributions (Figures 6.4–6.8)

GLOS surveyors recorded the locations (including distances and compass directions from survey corners) of "witness trees" or "bearing trees" at half-mile and one-mile intervals that represented the corners of quarter-sections (1/4 mile^2) or sections (1 mile^2) of land within a township-range coordinate system wherein each township was a 36-square-mile land parcel. The bearing trees were identified to species, and their diameters were recorded. The GLOS resulted in a series of consecutive, nonoverlapping sample areas of uniform size, located along a systematic grid work of one-mile (1.6-km) intervals (Delcourt and Delcourt, 1996).

The tree species in Figures 6.4 through 6.8 were selected for this lab exercise from a total of 28 species recorded by the GLOS (Delcourt and Delcourt, 1996). These trees were dominant within pre-European settlement forests and are characteristic of differing soil conditions across the study area. On the distribution maps, each dot represents a surveyed bearing tree; longer lines on some of the dot maps show surveyed distances described by the surveyors to contain continuous tracts of a particular tree type. White pine (Figure 6.4) was the largest and longest-lived species of pine native to the study area. It was the principal timber tree in early settlement times, originally occupying

a b

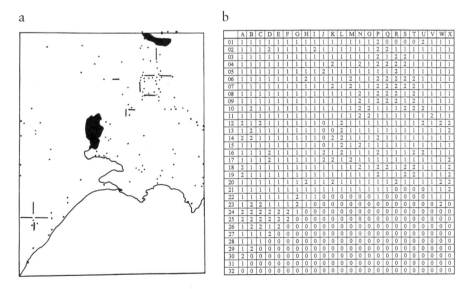

FIGURE 6.4
(a) Pre-European settlement distribution map for white pine (*Pinus strobus*), a xeric tree species. Each dot on the map represents a witness tree or bearing tree recorded in the U.S. General Land Office (GLO) Surveys of 1840–1849 (Delcourt and Delcourt, 1996).
(b) Raster grid derived from Figure 6.4a using a resolution of 0.5 × 0.5 mile (0.25 mile^2). The FRAGSTATS-compatible (ASCII) file is designated **whitpine.fra** on the CD where 0 = no data; 1 = absent; 2 = present.

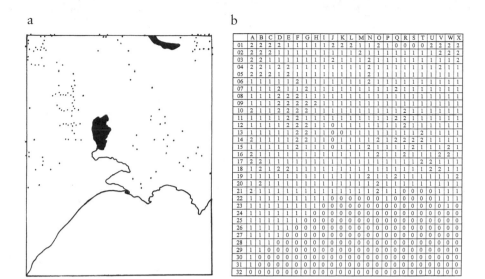

FIGURE 6.5
(a) Pre-European settlement distribution map for sugar maple (*Acer saccharum*), a mesic tree species. Each dot represents a witness tree or bearing tree recorded in the U.S. GLO Surveys of 1840–1849 (Delcourt and Delcourt, 1996).
(b) Raster grid derived from Figure 6.5a using a resolution of 0.5 × 0.5 mile. The FRAGSTATS-compatible file is **sugmaple.fra** where 0 = no data; 1 = absent; 2 = present.

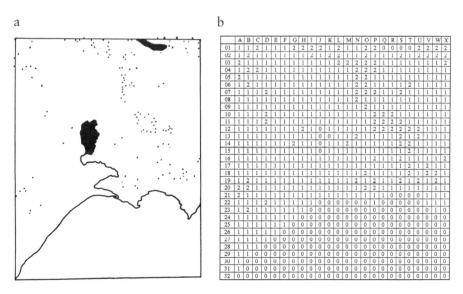

FIGURE 6.6
(a) Pre-European settlement distribution map for American beech (*Fagus grandifolia*), a mesic tree species. Each dot represents a witness tree or bearing tree recorded in the U.S. GLO Surveys of 1840–1849 (Delcourt and Delcourt, 1996).
(b) Raster grid derived from Figure 6.6a using a resolution of 0.5 × 0.5 mile. The FRAGSTATS-compatible file is **ambeech.fra** where 0 = no data; 1 = absent; 2 = present.

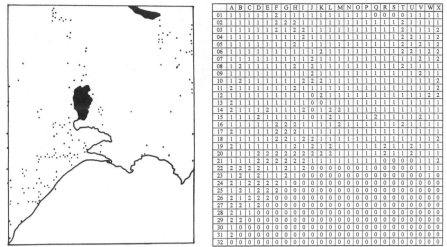

FIGURE 6.7
(a) Pre-European settlement distribution map for tamarack (*Larix laricina*), a hydric tree species. Each dot represents a witness tree or bearing tree recorded in the U.S. GLO Surveys of 1840–1849 (Delcourt and Delcourt, 1996).
(b) Raster grid derived from Figure 6.7a using a resolution of 0.5 × 0.5 mile. The FRAGSTATS-compatible file is **tamarack.fra** where 0 = no data; 1 = absent; 2 = present.

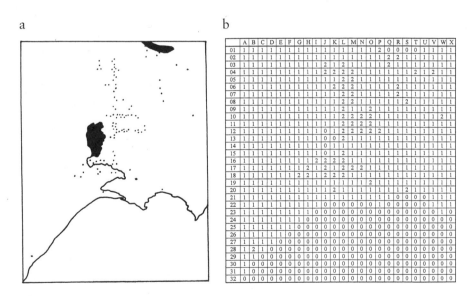

FIGURE 6.8
(a) Pre-European settlement distribution map for jack pine (*Pinus banksiana*), a fire disturbance–adapted, xeric tree species. Each dot represents a witness tree or bearing tree recorded in the U.S. GLO Surveys of 1840–1849 (Delcourt and Delcourt, 1996).
(b) Raster grid derived from Figure 6.8a using a resolution of 0.5 × 0.5 mile. The FRAGSTATS-compatible file is **jackpine.fra** where 0 = no data; 1 = absent; 2 = present.

extensive tracts of land and attaining up to 60% of the forest composition in the northeastern portion of the study area. Sugar maple (Figure 6.5) is a mesic tree species that in pre-European settlement times reached up to 60% of the forest composition in the northwestern quadrant of the study area, with smaller populations in the northeastern quadrant. In contrast to sugar maple, a second mesic tree species, American beech (Figure 6.6) was not abundant west of Lake Millecoquins but was important along a generally northwest-to-southeast trending line extending through the central part of the study area. Tamarack (Figure 6.7) was an important wetland tree, occurring along stream drainages and in wet hollows between sand ridges along ancient shorelines of Lake Michigan. Isolated tamarack trees also grew in open wetlands. Jack pine (Figure 6.8) is a fire-adapted species characteristically found growing on open upland sand plains. The serotinous cones of jack pine trees require the intense heat of wildfires to open and disperse their seeds.

Pre-European Settlement Disturbance Regime

The GLOS plat maps include locations of both natural and human-caused disturbances, including tree windfalls and burned areas as well as forest openings and meadows made by Native Americans near their villages and maple sugar camps. Two large windfalls were recorded in the 1840–1849 survey of the study area. These were blowdowns of timber located in cedar swamps west of Lake Millecoquins and along the wind-exposed shore of Lake Michigan near the present site of Naubinway.

Two types of burned areas were recognized: (1) wildfires located in the pinelands northeast of Lake Millecoquins and in the southwestern portion of the study area along crests of ancient beach ridges on the coastal lake plain, and (2) burned areas associated with an Ottawa village located on the southern shore of Lake Millecoquins (Figures 6.3 and 6.9). Tree stubs and standing dead trees were mapped within the burned area in the southwestern part of the study area and possibly reflect a primary disturbance in the form of an outbreak of spruce budworm that then provided the fuel source for a wildfire event (Delcourt and Delcourt, 1996).

Post-Settlement Land-Use History (Figure 6.10)

Original forests were cut locally beginning after 1870 (White and Mladenoff, 1994), with old-growth white pine groves identified by the GLOS reports selectively removed first. Intense slash fires followed historic logging of the white pine. By the end of the 19th century, extensive tracts of hardwoods and cedar were logged for furniture, cordwood, railroad ties, and telegraph poles.

West of Lake Millecoquins, much of the land originally covered with mesic upland hardwood forest was converted to pasture for beef and dairy cattle. During the European settlement, logging, and conversion of the land, railroad lines were constructed linking the Upper Peninsula of Michigan with markets in Minnesota, Wisconsin, and Ontario (White and Mladenoff, 1994). Small villages such as Engadine, Michigan, were established as water stops to ser-

a   b

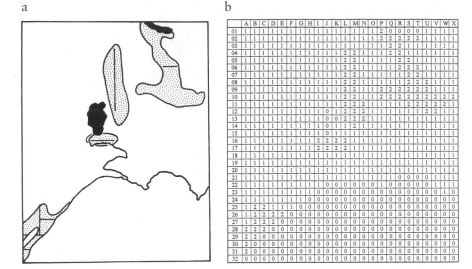

	A	B	C	D	E	F	G	H	I	J	K	L	M	N	O	P	Q	R	S	T	U	V	W	X
01	1	1	1	1	1	1	1	1	1	1	1	1	1	1	1	2	0	0	0	0	1	1	1	1
02	1	1	1	1	1	1	1	1	1	1	1	1	1	1	1	2	2	2	2	2	1	1	1	1
03	1	1	1	1	1	1	1	1	1	1	1	1	1	1	1	2	2	1	1	1	1	1	1	1
04	1	1	1	1	1	1	1	1	1	1	1	2	2	1	1	1	2	2	1	1	1	1	1	1
05	1	1	1	1	1	1	1	1	1	1	1	2	2	1	1	1	2	2	1	1	1	1	1	1
06	1	1	1	1	1	1	1	1	1	1	1	2	2	1	1	1	2	2	2	1	1	1	1	1
07	1	1	1	1	1	1	1	1	1	1	1	2	2	1	1	1	1	2	2	1	1	1	1	1
08	1	1	1	1	1	1	1	1	1	1	1	2	2	1	1	1	1	2	2	2	1	1	1	1
09	1	1	1	1	1	1	1	1	1	1	1	2	2	1	1	2	2	2	2	2	1	1	1	1
10	1	1	1	1	1	1	1	1	1	1	1	2	2	1	1	2	2	2	2	2	2	2	2	2
11	1	1	1	1	1	1	1	1	1	1	1	1	2	2	1	1	1	1	2	2	2	2	2	1
12	1	1	1	1	1	1	1	1	1	1	0	1	2	2	2	1	1	1	1	1	2	2	1	1
13	1	1	1	1	1	1	1	1	1	1	0	0	2	2	2	1	1	1	1	1	1	1	1	1
14	1	1	1	1	1	1	1	1	1	1	0	1	1	2	1	1	1	1	1	1	1	1	1	1
15	1	1	1	1	1	1	1	1	1	1	0	1	1	1	1	1	1	1	1	1	1	1	1	1
16	1	1	1	1	1	1	1	1	1	1	2	2	2	2	1	1	1	1	1	1	1	1	1	1
17	1	1	1	1	1	1	1	1	1	1	2	2	2	2	1	1	1	1	1	1	1	1	1	1
18	1	1	1	1	1	1	1	1	1	1	1	1	1	1	1	1	1	1	1	1	1	1	1	1
19	1	1	1	1	1	1	1	1	1	1	1	1	1	1	1	1	1	1	1	1	1	1	1	1
20	1	1	1	1	1	1	1	1	1	1	1	1	1	1	1	1	1	1	1	1	1	1	1	1
21	1	1	1	1	1	1	1	1	1	1	1	1	1	1	1	1	1	0	0	0	0	1	1	1
22	1	1	1	1	1	1	1	1	1	0	0	0	0	0	0	0	1	0	0	0	0	1	1	1
23	1	1	1	1	1	1	1	1	0	0	0	0	0	0	0	0	0	0	0	0	0	1	1	0
24	1	1	1	1	1	1	1	1	0	0	0	0	0	0	0	0	0	0	0	0	0	0	0	0
25	1	2	2	1	1	1	1	0	0	0	0	0	0	0	0	0	0	0	0	0	0	0	0	0
26	1	2	2	2	2	0	0	0	0	0	0	0	0	0	0	0	0	0	0	0	0	0	0	0
27	1	2	2	2	0	0	0	0	0	0	0	0	0	0	0	0	0	0	0	0	0	0	0	0
28	2	2	2	0	0	0	0	0	0	0	0	0	0	0	0	0	0	0	0	0	0	0	0	0
29	2	2	0	0	0	0	0	0	0	0	0	0	0	0	0	0	0	0	0	0	0	0	0	0
30	2	0	0	0	0	0	0	0	0	0	0	0	0	0	0	0	0	0	0	0	0	0	0	0
31	2	0	0	0	0	0	0	0	0	0	0	0	0	0	0	0	0	0	0	0	0	0	0	0
32	0	0	0	0	0	0	0	0	0	0	0	0	0	0	0	0	0	0	0	0	0	0	0	0

FIGURE 6.9
(a) Pre-European settlement fire regime map based on field notes and plat maps of the U.S. GLO Surveys of 1840–1849 and including both wildfires and Native American–set fires (Delcourt and Delcourt, 1996). Burned areas are shown as patterned dots.
(b) Raster grid derived from Figure 6.9a using a resolution of 0.5 × 0.5 mile. The FRAGSTATS-compatible file is **fire.fra** where 0 = no data; 1 = unburned; 2 = burned.

vice steam-driven locomotives. Lumber mills and roof-shingle mills were located initially in coastal villages such as Naubinway and later along railways farther inland, in Engadine and Newberry (Figure 6.1). At Newberry, industrial facilities were developed to render hardwood trees into charcoal fuel for the iron industry and distillation chemicals such as methanol. Similar processing occurred across the northern conifer–mixed hardwoods region from Maine to Minnesota. By the early 1900s, even the once-vast cedar swamps were cut for their decay-resistant wood, which was used for railroad ties, shingles, and fenceposts.

In the 1920s, private landowners purchased a tract of cut-over land encompassing 62-square-mile sections located north of Naubinway and east of Lake Millecoquins. With the establishment of the Hiawatha Sportsman's Club (HSC) in 1927, the modern era of fire suppression, reforestation, and stewardship began. In order to encourage increases in the deer population, early successional forests of aspen and birch were allowed to regenerate. Sites formerly occupied by large white pine trees were planted in red and white pine seedlings where possible, but some stump fields whose soil had burned so intensely that regeneration was hindered became permanent forest openings and blueberry "barrens." As the proportion of forest to field decreased and the

a   b

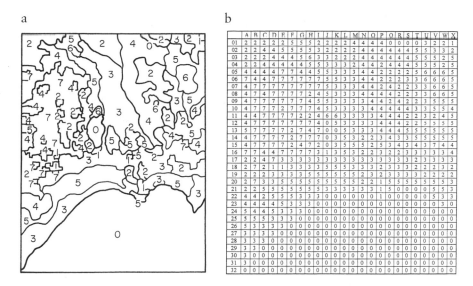

FIGURE 6.10
(a) Late-20th-century land-use map, taken from the Open File Land Use series of maps of land use and land cover, Sault Ste. Marie, Michigan, 1:250,000 scale, minimum resolution of four hectares (U.S. Geological Survey, LUDA series, 1980). Cover types are designated as follows:
0 = lake;
1 = cultural feature (urban, built-up land, or quarry);
2 = upland deciduous forest;
3 = upland evergreen forest;
4 = mixed deciduous/evergreen forest;
5 = forested wetland;
6 = nonforested wetland;
7 = cropland or pasture.
(b) Raster grid derived from Figure 6.10a using a resolution of 0.5 × 0.5 mile (0.25 mile^2); cover types, assigned using 50% cover rule. The FRAGSTATS-compatible (ASCII) file is designated **post.fra** on the CD.

extent of forest edge increased, populations of white-tailed deer multiplied (White and Mladenoff, 1994).

Today, the HSC manages its land to promote early successional forests as deer habitat and to preserve the few remaining groves of old-growth hardwoods. On surrounding state-owned lands maintained by the Michigan Department of Natural Resources, forest compartments are managed to (1) maintain the remaining tracts of wetlands, (2) allow forests to regenerate to old-growth status along a corridor adjacent to Lake Michigan that serves as habitat for forest-interior species such as moose and bobcat, and (3) continue to use areas of successional hardwoods for pulpwood to support paper and particle-board manufacturing industries.

EXERCISES

This lab includes a *basic option* and an *advanced option*. The basic option can be completed "by-hand," whereas the advanced option includes FRAGSTATS analyses. To complete either exercise in one lab period, students may work in teams on different parts of the exercise, then pool their results before answering questions. Alternatively, the exercise may be assigned over two separate lab periods. In addition, the instructor may want to prepare color-enhanced maps and transparencies ahead of time to make more efficient use of lab time.

PART I. BASIC OPTION

EXERCISE 1.1
Analyzing Landscape Pattern: A Comparison of Pre- and Post-Settlement Landscape Patterns

In this exercise you will analyze raster maps and calculate a suite of landscape metrics to compare landscape patterns between the pre-European settlement and post-settlement landscapes.

1. Color-code the different cover types on each of the following raster maps:
   (a) Pre-European settlement vegetation (Figure 6.3b)
   (b) Late 20th century land use (Figure 6.10b)
   Use a different color for each cover type, but use the same colors for comparable cover types of pre-European settlement vegetation and late 20th century land use. Be sure to print out and use the larger version of these maps that have been provided on the CD as pdf files, which can be viewed using Adobe Acrobat (available freely on the web).

2. Identify distinct patches of the different cover types using the four-neighbor rule. In other words, a patch includes only those cells that touch along the flat edge of a cell and excludes cells that touch diagonally, at a corner.

3. For both raster maps, count the number of cells in each discrete patch. Multiply the number of cells in each patch by 65 to convert to hectares. For both of these raster maps, calculate the following indices for patchiness of the landscape (be sure to exclude cells located in Lake Michigan from the analysis):
   (a) Total number of patches for each cover type
   (b) Total area of each cover type, summed over all its patches
   (c) Area of the largest patch in each cover type
   (d) Mean patch size for each cover type

   Record your pre-European settlement map results in Table 6.2 and your post-settlement results in Table 6.3, which can be printed from the CD.

Question 1.1. On the pre-European settlement vegetation map, which cover type had the largest total area? Did this cover type completely surround other, smaller cover types as a matrix? Did this cover type occur as a single, large patch or as numerous smaller patches? Was this cover type broadly distributed across the study area, or was it confined to a particular geographic sector? On which soil type(s) did this cover type occur?

Question 1.2. Overall, was a higher proportion of the pre-European settlement landscape occupied by upland forests or by wetlands?

Question 1.3. Were all pre-European settlement cover types similar in the number of patches and their mean sizes, or did they differ widely in their landscape-level metrics?

Question 1.4. What has been the effect of European American settlement on the openness and patchiness of the landscape? To answer this question, compare the total number of patches, the size of the largest patch, and the mean patch size on the pre-European settlement vegetation map with those on the map of late 20th century land use for the following:

(a) Cover type 1 (cultural features)

(b) Cover type 7 (Native American meadow versus cropland or pasture)

Question 1.5. How have human activities over the past 150 years affected the size, distribution, and contiguity of wetlands? (Combine cover types 5 and 6, and compare on the two maps.)

Question 1.6. What changes have taken place in the past 150 years in the relative proportions and patchiness of upland deciduous, upland evergreen, and mixed deciduous/evergreen forests? (Compare cover types 2, 3, and 4 on the two maps.)

EXERCISE 1.2
Determining Causes of Landscape Heterogeneity:
Soil Factors and Pre-European Settlement Tree Species

Next, you will examine the factors governing the Pre-European settlement distribution of several tree species by creating a series of map overlays and summary charts.

1. Make six transparencies of the blank grid map (Figure 6.11) to use as overlays. These can be printed from the CD.

2. For each tree distribution map (Figures 6.4b–6.8b), overlay a transparency of a blank grid map and highlight the grid squares occupied by the species (cover type 2).

3. Color-code the cover types on the soils raster map (Figure 6.2b).

4. Next, compare each tree distribution in turn with the soils raster map. Count the number of grid cells for each species on each soil type in which it occurs and multiply the number of cells by 65 to convert to hectares. Enter the area in Table 6.4 on the CD.

FIGURE 6.11
Blank tally sheet with grid resolution of 0.5 × 0.5 mile (0.25 mile^2). Cover type 0 = lake. A larger version of this map is available on the CD as a pdf file.

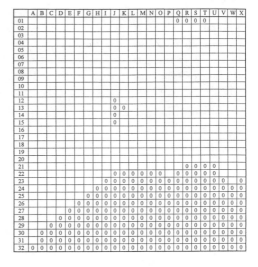

5. Using a separate piece of graph paper for each species, plot a bar graph showing the area of each soil type occupied by the tree species.

EXERCISE 1.3
Determining Causes of Landscape Heterogeneity:
Disturbance Regimes and Pre-European Settlement Distribution of Jack Pine

Here you will examine the distribution of jack pine relative to past fires.

1. On a transparency of the fire disturbance map (Figure 6.9b), high-light the grid squares lying within the burned areas (patches of cover type 2).

2. Using your newly created fire map and your transparency for jack pine (from Figure 6.8b) that you created in Exercise 1.2, tally the number of grid cells for each of the following cases:
   (a) Both jack pine and fire are present
   (b) Both jack pine and fire are absent
   (c) Jack pine present, fire absent
   (d) Jack pine absent, fire present
   Of course, only include cells on land. Convert your numbers to measurements in hectares and enter in Table 6.5 from the CD.

3. Compare the distribution of jack pine with that of fire disturbance by plotting the area in which both jack and fire disturbance occurred versus the number or area of grid cells in which either jack pine, fire disturbance, or both were absent.

Question 1.7. Which tree species are most strongly associated with xeric, acidic, sandy soils without ortstein layers (soil cover type 1)? What adaptations do these species share that may help explain their overall distribution on this very well drained and nutrient-poor soil?

Question 1.8. Which species are found most commonly on sandy soils with ortstein layers or on mesic, calcareous soils (cover types 2 and 3)? In what general vegetation cover types were these tree species dominant in pre-European settlement times? (Refer to Figure 6.3a).

Question 1.9. How closely does the pre-European settlement distribution of tamarack follow the poorly drained habitat corresponding with soil cover type 6 (hydric, acidic sand and mucky peat)? Is tamarack found exclusively on this soil type, or does it extend to any other soil types?

Question 1.10. How well does the presence of jack pine circa 1850 indicate the distribution of burned sites mapped in the GLOS? Was jack pine found only within fire-disturbed areas? If not, what other site factors may explain its distribution?

II. ADVANCED OPTION

EXERCISE 2.1
Analyzing Landscape Pattern:
A Comparison of Pre- and Post-Settlement Landscape Patterns

In this exercise you will use FRAGSTATS to analyze raster maps and calculate a set of landscape metrics to compare landscape patterns between the pre-European settlement and post-settlement landscapes. The necessary files, the FRAGSTATS-compatible (*.fra) files and the PC version of FRAGSTATS (McGarigal and Marks, 1995) are provided on the CD under the directory for this lab. It is assumed you are familiar with the directions and documentation for FRAGSTATS that can be found in Chapter 7, Understanding Landscape Metrics I, and on the CD for Chapter 7.

For this exercise, the following input to FRAGSTATS will be helpful:

1. The file **pre.fra** represents a grid of the pre-European settlement vegetation, and **post.fra** is a grid of the 20th-century land-use maps. Both files are formatted as ASCII text for input to FRAGSTATS.

2. The landscapes are 32 rows × 24 columns.

3. Each grid cell is 805 meters on a side.

4. Background values must be designated as 9 in order to exclude cells representing Lake Michigan from the analysis (marked −9 on the data files but as zeroes in the figures in the text).

Use FRAGSTATS to perform the following analyses for both pre-European settlement vegetation and late-20th-century land use:

1. Calculate the following metrics of spatial configuration:
   (a) Number of patches for each cover type
   (b) Total area of each cover type
   (c) Area of largest patch of each cover type
   (d) Mean patch size for all cover types
   Use Tables 6.2 and 6.3 from the CD to summarize your results.

2. Calculate the following metrics of landscape composition:
   (a) Proportions (p_i) occupied by each cover type
   (b) Shannon Evenness diversity index. Record your results in Table
       6.6 from the CD.

3. Also for Table 6.6, sum the appropriate cover types to calculate the
   total area in:
   (a) Forest
   (b) Nonforest
   (c) Upland vegetation
   (d) Wetland vegetation

Answer *Questions 1.1 through 1.6* from the **basic option** in addition to the
following questions.

Question 2.1. Consider Table 6.6. Which of the cover types in pre-
European settlement times occupied the greatest proportion of the land in the
study area? Did this cover type still occupy the highest proportion of the land
in the late 20th century? How did the proportion of forested land versus nonforested land change over the past 150 years? How did the proportion of upland vegetation versus wetland vegetation change?

Question 2.2. Does the value obtained for the Shannon evenness index for
pre-European settlement vegetation indicate that this landscape was highly diverse or relatively homogeneous? How has that diversity changed in the past
150 years?

EXERCISE 2.2
Determing Causes of Landscape Heterogeneity:
Soil Factors and Pre-Settlement Distribution of American Beech

Here, you will examine the relationship between the distribution of American beech (Figure 6.6b) and the distribution of xeric, acidic, sandy soils
with ortstein layers (soil cover type 2 in Figure 6.2b).

1. Overlay the transparency of Figure 6.6b on Figure 6.2b and count
   the number of cells (on land) in which
   (a) both American beech and soil cover type 2 are present;
   (b) both American beech and soil cover type 2 are absent;
   (c) American beech is present but soil cover type 2 is absent;
   (d) American beech is absent but soil cover type 2 is present.

TABLE 6.7
CHI-SQUARE CALCULATIONS

(a) Calculation of chi-square test of association for american beech and soil cover
    type 2

$X^2 = \text{sum}_i (\text{observed}_i - \text{expected}_i)^2 / \text{expected}_i$

$a = (a + b)(a + c) / n$
$b = (a + b)(b + d) / n$
$c = (c + d)(a + c) / n$
$d = (c + d)(b + d) / n$

	Observed Frequencies			Expected Frequencies	

Soil type 2	Beech present		Soil type 2	Beech present	
	Yes	No		Yes	No
Yes			Yes		
No			No		

$X^2 = \boxed{}$

Critical $X^2 = 10.83$ with 1 degree of freedom at $P = 0.0001$.
H_o = American beech is not associated with the soil cover type 2.
Reject H_o if calculated X^2 is > 10.83.

(b) Calculation of chi-square test of association for jack pine and fire disturbance

	Observed Frequencies			Expected Frequencies	
Fire disturb. present	Jack pine present		Fire disturb. present	Jack pine present	
	Yes	No		Yes	No
Yes			Yes		
No			No		

$X^2 = \boxed{}$

Critical $X^2 = 10.83$ with 1 degree of freedom at $P = 0.0001$.
H_o = Jack pine is not associated with the presence of fire disturbance.
Reject H_o if calculated X^2 is > 10.83.

(c) Calculation of chi square test of association for jack pine and soil type 1

	Observed Frequencies			Expected Frequencies	
Soil type 1	Jack pine present		Soil type 1	Jack pine present	
	Yes	No		Yes	No
Yes			Yes		
No			No		

$X^2 = \boxed{}$

Critical $X^2 = 10.83$ with 1 degree of freedom at $P = 0.0001$.
H_o = Jack pine is not associated with the soil cover type 1.
Reject H_o if calculated X^2 is > 10.83.

TABLE 6.8
EXAMPLE METHOD FOR CALCULATING χ^2 TEST OF ASSOCIATION
(INDEPENDENCE) BETWEEN TREE DISTRIBUTIONS (AMERICAN BEECH)
AND PHYSICAL ENVIRONMENT (SOIL COVER TYPE 2) AS IN TABLE 6.7A.
CALCULATE OBSERVED FREQUENCIES AS SHOWN IN PART (A) BY FILLING
IN CELLS A, B, C, AND D WITH YOUR MEASUREMENTS MADE FROM MAPS.
THEN, CALCULATE EXPECTED FREQUENCIES AS SHOWN IN PART (B), US-
ING THE VALUES OF A, B, C, AND D FROM THE OBSERVED FREQUENCIES
TABLE. CALCULATE THE χ^2 VALUE AS SHOWN IN (C) AND COMPARE TO
CRITICAL LEVEL OF χ^2.

(a)

Observed (Measured) Frequencies

		Tree Present		
		Yes	No	Total
Soil cover	Yes	a	b	$a + b$
type present	No	c	d	$c + d$
	Total	$a + c$	$b + d$	n

(b)

Expected Frequencies

		Tree Present	
		Yes	No
Soil cover	Yes	$(a + b)(a + c) / n$	$(a + b)(b + d) / n$
type present	No	$(c + d)(a + c) / n$	$(c + d)(b + d) / n$

(c)

$$\chi^2 = \sum_i \frac{(observed_i - \text{expected}_i)^2}{\text{expected}_i}$$

Critical $\chi^2 = 10.83$ with 1 degree of freedom at $p = 0.001$.
H_o = American beech is not associated with soil cover type 2.
Reject H_o if calculated χ^2 is >10.83.

2. These numbers are your *observed* frequencies. Enter your results in the appropriate squares in Table 6.7a.

3. Examine Table 6.8, which explains the calculations required for the χ^2 test of association, and be sure you understand the procedure.

4. Next, perform a χ^2 test of association between American beech and soil type 2, either by hand as described below, or using the spreadsheet (**chisquare.xls**). Given your input data of observed data, this spreadsheet will automatically calculate your expected frequencies and χ^2 values.

    (a) First, calculate the *expected* frequencies using your observed data and the appropriate equations from Table 6.8. Enter your results in Table 6.7.

    (b) Then, calculate the χ^2 value and determine the significance level.

EXERCISE 2.3
Determining Causes of Landscape Heterogeneity:
Disturbance Regimes, Soil Types, and the Pre-European
Settlement Distribution of Jack Pine

1. Here you will calculate the strength of the relationship between the distribution of jack pine (Figure 6.8b) and fire disturbance (Figure 6.9b) by overlaying the transparency of Figure 6.8b on Figure 6.9b.

2. Then, calculate the χ^2 statistic as in Exercise 2.2. Report your data in Table 6.7b, or use the spreadsheet **chisquare.xls** for your calculations.

3. Repeat the chi-square analysis for jack pine versus xeric, acidic, sandy soils without ortstein (soil type 1 on Figure 6.2b).

Question 2.3. How strong is the association between American beech and sandy soils with a strongly developed ortstein layer? On what additional soil types does American beech occur? What types of soils are not favorable for populations of American beech?

Question 2.4. Was the pre-European settlement distribution of jack pine more strongly associated with fire disturbance or with the distribution of soil type 1? On what additional soil types did jack pine occur? If the distribution of soil type 1 was overlaid with the distribution of fire disturbance, do you think the correlation of jack pine distribution with the combination of soil cover type 1 plus fire would be improved or weakened? Why?

BIBLIOGRAPHY

Note. An asterisk preceding the entry indicates that it is a suggested reading.

DELCOURT, H. R., AND P. A. DELCOURT. 1996. Presettlement landscape heterogeneity: Evaluating grain of resolution using General Land Office Survey data. *Landscape Ecology* 11:363–381.

*DELCOURT, H. R., AND P. A. DELCOURT. 2000. Eastern deciduous forests. In M. G. Barbour and D. W. Billings, eds. *North American Terrestrial Vegetation*, 2nd ed. Cambridge University Press, Cambridge, UK, chapter 10. A discussion of the environmental factors that influence the distributions of tree species important within the mixed conifer–northern hardwoods forest.

DELCOURT, P. A., W. H. PETTY, AND H. R. DELCOURT. 1996. Late-Holocene formation of Lake Michigan beach ridges correlated with a 70-yr oscillation in global climate. *Quaternary Research* 45:321–326.

*FLADER, S. L., ED. *The Great Lakes Forest: An Environmental and Social History.* University of Minnesota Press, Minneapolis. The chapter by E. A. Bourdo Jr., pp. 3–16, entitled "The Forest the Settlers Saw," gives insight into the methods and observations of the U.S. General Land Office Surveyors, as well as the variation in composition of the pre-European settlement forests of Upper Michigan.

MCGARIGAL, K., AND B. J. MARKS. 1995. FRAGSTATS: Spatial pattern analysis program for quantifying landscape structure. *General Technical Report PNW-GTR-351*, U.S. Department of Agriculture, Forest Service, Pacific Northwest Research Station, Portland, Oregon.

PETTY, W. H., P. A. DELCOURT, AND H. R. DELCOURT. 1996. Holocene lake-level fluctuations and beach-ridge development along the northern shore of Lake Michigan, U.S.A. *Journal of Paleolimnology* 15:147–169.

SILBERNAGEL, J., S. R. MARTIN, M. R. GALE, AND J. CHEN. 1997. Prehistoric, historic, and present settlement patterns related to ecological hierarchy in the eastern Upper Peninsula of Michigan, U.S.A. *Landscape Ecology* 12:223–240.

UNITED STATES GEOLOGICAL SURVEY, Open File Map 84-035-1, Land Use Series Land Use and Land Cover, 1980, Sault Sainte Marie, Michigan, U.S.; Ontario, Canada (U.S. Portion Only), Scale 1:250,000. Available from the National Cartographic Information Center, USGS, Reston, Virginia 22092.

WHITE, M. A., AND D. J. MLADENOFF. 1994. Old-growth forest landscape transitions from pre-European settlement to present. *Landscape Ecology* 9:191–205.

*WHITNEY, G. G. 1994. *From Coastal Wilderness to Fruited Plain, a History of Environmental Change in Temperate North America, 1500 to the Present.* Cambridge University Press, Cambridge, UK. A comprehensive overview of the historic transformation of American forests.

WHITNEY, G., S. ROPACK, AND C. OUTWATER. 1995. Soil survey of Mackinac County, Michigan. U.S.D.A. Soil Conservation Service, U.S. Government Printing Office, Washington, D.C.

QUANTIFYING LANDSCAPE PATTERN

Quantifying landscape pattern is necessary to understanding the relationships between pattern and process. Landscape ecology employs a variety of methods for the detection and description of landscape pattern. Chapter 7, Understanding Landscape Metrics I, introduces different types of landscape metrics and explains the calculation and rationale behind several that are commonly used. Furthermore, the lab provides experience using FRAGSTATS, one of the most widely used programs for pattern analysis. Understanding the influence of scale on landscape metrics is fundamental to using and interpreting changes in landscape metrics appropriately. Chapter 8, Understanding Landscape Metrics II: Effects of Changes in Scale, builds directly on Chapter 7 by examining the influence of changes in grain, extent, and classification scheme on landscape metrics. Chapter 9, Neutral Landscape Models, explores patch delineation and the creation of neutral landscape models, using the program Rule. Neutral landscape models have been fundamental to exploring landscape change, the spread of disturbance, and the movement of organisms across landscapes. Chapter 10, Scale Detection Using Semivariograms and Autocorrelograms, presents an accessible (using only spreadsheet software!) treatment of the use of semivariograms and autocorrelograms for the detection of the scale of landscape variability.

CHAPTER 7

Understanding Landscape Metrics I

Jeffrey A. Cardille and Monica G. Turner

OBJECTIVES

An extensive set of landscape metrics exists to quantify spatial patterns in heterogeneous landscapes. Developers and users of these metrics typically seek to *objectively* describe landscapes that humans assess *subjectively* as "clumpy," "dispersed," "random," "diverse," or "fragmented" for example. Because the quantification of pattern is fundamental to many of the relationships we seek to understand in landscape ecology, a basic familiarity with the most commonly used metrics is extremely important. While several software programs (e.g., Fragstats [McGarigal and Marks, 1995] and r.le [Baker and Cai, 1992] in the Grass geographic information system) evaluate maps quickly and cheaply, there are no absolute rules governing the proper use of landscape metrics. Thus, in this lab, to help foster the appropriate use of landscape metrics, students will

1. gain familiarity with some commonly used metrics of landscape pattern;
2. distinguish metrics that describe landscape composition from those that describe spatial configuration; and
3. understand some of the factors that influence the selection and interpretation of landscape metrics.

In addition, computer-based sections of the lab (Parts 3 and 4) are provided to meet the following objectives:

1. provide experience with landscape pattern analysis using FRAGSTATS; and
2. illustrate the correlation structure among some commonly used landscape metrics.

In this lab, a set of commonly used metrics in landscape ecology is presented. Emphasis is placed on the understanding gained from actually calculating select metrics by hand (using small landscapes and a calculator) rather than using only a metric-calculation package. Although the landscapes used for the hand calculations are much smaller than those typically input to metric-calculation software packages, the concepts and equations learned in this lab are the same as those used for full-sized images. Sections using FRAGSTATS and actual landscape images are also included to investigate the behavior of landscape metrics in a more realistic setting. The CD contains digital versions (.pdf files) of all the images and figures you will use in this lab, under the directory **"Images"** for this chapter. These images can be viewed using Adobe Acrobat Reader, which is available free on the Web. The CD also contains a pdf file of all the data entry tables, which you should print before starting the lab.

Note regarding Fragstats 2.0 and Windows XP

The version of Fragstats included on the CD (FRAGSTATS 2.0) is not compatible with Windows XP. Using Fragstats on Windows XP requires downloading the latest shareware version of Fragstats 3.0 from the web at
http://www.umass.edu/landeco/research/fragstats/fragstats.html

INTRODUCTION

The quantification of landscape pattern has received considerable attention since the early 1980s, in terms of both development and application (Romme and Knight, 1982; O'Neill et al., 1988; Turner et al., 1989; Baker and Cai, 1992; Wickham and Norton, 1994; Haines-Young and Chopping, 1996; Gustafson, 1998). Several of the most commonly used landscape metrics were derived originally from percolation theory, fractal geometry, and information theory (the same branch of mathematics that led to the development of species diversity indices). The increased availability of spatial data, particularly over the past two decades, has also presented myriad opportunities for the development, testing, and application of landscape metrics.

Why are methods for describing and quantifying spatial pattern such necessary tools in landscape ecology? Because landscape ecology emphasizes the interactions among spatial patterns and ecological processes, one needs to understand and quantify the landscape pattern in order to relate it to a process. Practical applications of pattern quantification include describing how a landscape has changed through time; making future predictions regarding landscape change; determining whether patterns on two or more landscapes differ from one another, and in what ways; evaluating alternative land management strategies in terms of the landscape patterns that may result; and determining whether a particular spatial pattern is conducive to movement by a particular organism, the spread of disturbance, or the redistribution of nutrients. In all of these cases, the calculation of landscape metrics is necessary to rigorously describe landscape patterns. However, relating these metrics of pattern to dynamic ecological processes still remains an area in need of further research.

Next, you will examine and manually calculate several commonly used landscape metrics using a calculator and a small landscape to ensure that you understand their underlying mathematics (Parts 1 and 2). Then, once you have a basic understanding of these metrics, two computer-based exercises (Parts 3 and 4) are provided to allow you to calculate metrics using FRAGSTATS and larger landscape images.

EXERCISE

Part 1. Metrics of Landscape Composition

The simplest landscape metrics focus on the composition of a landscape (e.g., which categories are present and how much of the categories there are), ignoring the specific spatial arrangement of the categories on the landscape. In this section, you will examine three metrics designed to assess the composition of a landscape: (1) the proportion of the landscape occupied by each cover type, (2) dominance, and (3) Shannon evenness.

Proportion (p_i) of the landscape occupied by the ith cover type is the most fundamental metric and is calculated as follows:

$$p_i = \frac{\text{Total number of cells of category } i}{\text{Total number of cells in the landscape}}$$

Proportions of different landscape types have a strong influence on other aspects of pattern, such as patch size or length of edge in the landscape (Gardner et al., 1987; Gustafson and Parker, 1992), and p_i values are used in the calculation of many other metrics. Several metrics derived from information theory use the p_i values of all cover types to compute one value that describes an entire landscape. Information theoretic metrics were first applied to landscape analyses by Romme (1982) to describe changes in the area occupied by forests of varying successional stage through time in a watershed in Yellowstone National Park, Wyoming (USA). Romme reasoned that the indices used to quantify species diversity in different communities could be modified and applied to describe the diversity of landscapes. Dominance and Shannon evenness are two such metrics that characterize how evenly the proportions of cover types occur within a landscape.

Dominance (D) (O'Neill et al., 1988) can be calculated as:

$$D = \frac{\ln(S) + \sum_i [p_i * \ln(p_i)]}{\ln(S)}$$

where S is the number of cover types, p_i is the proportion of the ith cover type, and ln is the natural log function. Values for D range between 0 and 1; values near 1 indicate a landscape dominated by one or few cover types, while values near 0 indicate that the proportions of each cover type are nearly equal.

Shannon evenness (*SHEI*) (Pielou, 1975) can be calculated as:

$$SHEI = \frac{-\sum_i [p_i * \ln(p_i)]}{\ln(S)}$$

where S is the number of cover types, p_i is the proportion of the ith cover type, and ln is the natural log function. Values for *SHEI* range between 0 and 1; values near 1 indicate that the proportions of each cover type are nearly equal, while values near 0 indicate a landscape dominated by one or few cover types.

A very important detail to note in the alternative formulations of information theoretic metrics is whether a particular formulation has been normalized to a common scale. Some early applications of dominance and Shannon evenness were not normalized to a common scale (e.g., O'Neill et al., 1988). The nonnormalized forms of these metrics are very sensitive to the number of cover types S in the landscapes, and thus comparisons among landscapes that differed in S were problematic. Normalizing a metric ensures that its values fall within a standardized range, such as from 0 to 1 (and not from 0 to 157, for example!). With D and *SHEI*, the normalization involves dividing the numerator by the maximum possible value of the index (ln S), as shown earlier.

CALCULATIONS

To understand these metrics and calculate them by hand within a reasonable time frame, you will calculate the metrics for two small hypothetical landscapes represented as 10×10 grids (Figure 7.1). It may be useful to print extra copies of the images on the CD-ROM, under the directory **"Images"** for this chapter, for your hand calculations.

Metrics of Landscape Composition in an Early-Settlement Landscape

An "early-settlement" landscape is shown on the left in Figure 7.1. This image represents an area that was previously fully forested, but has lost some forest to agricultural and urban uses. The landscape is composed of a 10×10 grid with each grid cell representing an area of one square kilometer (1000 m $\times$ 1000 m; 10^6 m^2).

Calculation 1.1: Calculate the proportions occupied by each of the three land covers in the early-settlement landscape. Record the values in Table 7.1 from the CD.

Calculation 1.2: Calculate dominance for the early-settlement landscape and record in Table 7.1.

Calculation 1.3: Calculate Shannon evenness for the early-settlement landscape and record in Table 7.1.

Simulated Landscapes

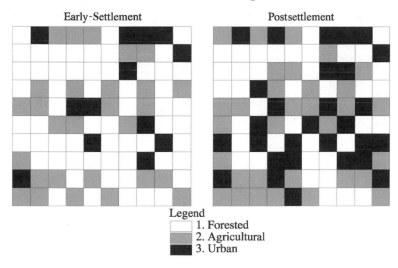

FIGURE 7.1
Hypothetical early-settlement and post-settlement landscape classifications.

Metrics of Landscape Composition in a Post-Settlement Landscape

A "post-settlement" landscape is shown on the right in Figure 7.1. This image represents the same area as the early-settlement landscape, but much later in time. Note that more of the forest has been converted to agricultural use. Additionally, some of the agricultural and forest land in the early-settlement image has been converted to urban use, while some of the early-settlement agricultural land has reverted to forest in the post-settlement image.

Calculation 1.4: Calculate the proportions occupied by each of the three land cover types in the post-settlement landscape. Record the values in Table 7.2 on the CD.

Calculation 1.5: Calculate dominance for the post-settlement landscape and record in Table 7.2.

Calculation 1.6: Calculate Shannon evenness for the post-settlement landscape and record in Table 7.2.

Considering the answers you obtained for both the early- and post-settlement landscapes, answer and discuss the following questions:

Question 1.1. How would you interpret/describe the changes in this landscape between the two time periods?

Question 1.2. Explain the relationship between dominance and Shannon evenness.

Question 1.3. If you were conducting an analysis of a real landscape, would you report both *D* and *SHEI*? Why or why not?

Question 1.4. To compare *D* or *SHEI* across two or more landscapes, does *S* need to be the same for each landscape in the comparison? Why or why not?

Question 1.5. Is there an upper and lower limit of *S* beyond which *D* and *SHEI* will not work?

Question 1.6. Use your calculator to perform some additional calculations of *D* assuming the proportions listed in Table 7.3 on the CD.

(a) Which of these hypothetical landscapes might be considered "similar" when only comparing *D*?

(b) Under what conditions could an interpretation of dominance (or other similar metrics) be problematic?

(c) Considering your interpretation of the data in Table 7.3, what other types of information and/or metrics would be necessary to distinguish these landscapes?

Part 2. Metrics of Spatial Configuration

A variety of landscape metrics are sensitive to the specific spatial arrangement of different cover types on a landscape. In this section, we will consider four components of landscape configuration: (1) patches, (2) edges, (3) probability of adjacency, and (4) contagion.

The **total number of patches** in a landscape results from first defining connected areas (i.e., patches or clusters) of each cover type *i*. Patches are commonly identified by using either of two neighbor rules. A patch may be identified using the **four-neighbor rule,** in which two grid cells are considered to be part of the same patch *only* if they are of the same cover type and share a flat adjacency between them. Alternatively, the **eight-neighbor rule** specifies that two grid cells of the same cover type are considered part of the same patch if they are adjacent *or diagonal* neighbors. In reporting the number of patches, it is important to distinguish whether the calculation is for all patches of all cover types or whether it is only for patches of a certain cover type *i*. In addition to the total number, patches can also be described in terms of their size (i.e., area) and edge-to-area ratio, which will be discussed later.

Mean patch size (MPS) is simply the arithmetic average size of each patch on the landscape or each patch of a given cover type. It is often calculated separately for each cover type as follows:

$$MPS = \frac{\sum_{k=1}^{m} A_k}{m}$$

where m = the number of patches for which the mean is being computed and A_k = the area of the kth patch. The units of area are defined by the user and should always be specified.

Edge calculations provide a useful measure of how dissected a spatial pattern is and can be calculated in a variety of ways. An edge is shared by two grid cells of different cover types when a side (not a corner) of one cell is adjacent to a side of the other cell (i.e., the four-neighbor rule). The total number of edges in a landscape can be calculated by counting the edges between different cover types for the entire landscape. Every edge in the landscape is counted only once.

Edge calculations are sometimes used to compute **edge-to-area ratios.** These may be computed in a variety of ways for a given landscape. For example, the total linear edge in a landscape can be divided by the area of the landscape to provide a single edge-to-area estimate, or edge density. More useful, however, are computations of edge-to-area ratios by cover type or for individual patches.

Edge calculations are sensitive to several factors. Whether the actual borders of the landscape image are considered as edges influences numbers of edges and edge-to-area ratios. (In this exercise, the landscape border will not be considered edge for your calculations.) Computer programs may use slightly different algorithms for totaling edges (e.g., the program RULE, used in Chapter 9, uses a different method for calculating total edge). It is extremely important to be consistent in both algorithm and units within a set of analyses. Additionally, although edge counts are relatively simple to compute from a landscape map, they are quite sensitive to the grain of the map (see Chapter 8, Understanding Landscape Metrics II: Effects of Changes in Scale).

Probability of adjacency $(q_{i,j})$ is the probability that a grid cell of cover type i is adjacent to a cell of cover type j. This metric is sensitive to the fine-scale spatial distribution of cover types and can be computed as:

$$q_{i,j} = \frac{n_{i,j}}{n_i}$$

where $n_{i,j}$ = the number of adjacencies between grid cells of cover type i and cover type j, and n_i = the total number of adjacencies for cover type i.

Probabilities of adjacency are often reported in an $S \times S$ matrix referred to as the **Q matrix.** Because they are probabilities, values for $q_{i,j}$ range from 0 to 1. High $q_{i,j}$ values indicate that the cells of cover type i have a high probability of being adjacent to cells of cover type j, while low $q_{i,j}$ values indicate a low probability. Values along the diagonals of the Q matrix (the $q_{i,i}$ values) are useful measures of the degree of clumping found *within* each cover type. High $q_{i,i}$ values indicate a highly aggregated, clumpy cover type, and low $q_{i,i}$ values indicate that the cover type tends to occur in isolated, dispersed grid cells or small patches.

The calculation of probabilities of adjacency may be performed in only the horizontal or only the vertical direction to detect directionality (referred to as anisotropy) in a pattern. For example, imagine a landscape composed of alternating ridges and valleys oriented in a north–south direction and in which forest cover occupies the ridges and agriculture occupies the valleys. The probabilities of adjacency would be different depending on whether you moved from north to south or from east to west across this landscape. In this lab, the horizontal and vertical values are averaged into a single measure of adjacency.

Contagion (C) (O'Neill et al., 1988; Li and Reynolds, 1993, 1994) uses the **Q matrix** values to compute an index of the overall degree of clumping in the landscape. Just as D and $SHEI$ used all p_i values for all cover types to compute one metric, contagion incorporates all $q_{i,j}$ values into one metric for the entire landscape. Contagion is useful in capturing relatively fine-scale differences in pattern that relate to the "texture" or "graininess" of the map. The equation is given by:

$$1 + \frac{\sum_i \sum_j [(p_i {}^* q_{i,j}){}^* \ln(p_i {}^* q_{i,j})]}{C_{\max}}$$

where $q_{i,j}$ = the adjacency probabilities defined above, and $C_{\max} = 2 {}^* \ln (S)$, which gives the maximum value of the index for a landscape with S cover types.

Values for contagion range from 0 to 1. A high contagion value indicates generally clumped patterns of landscape categories within the image, while values near 0 indicate a landscape with a dispersed pattern of landscape categories. Note that contagion can be computed differently if the $q_{i,j}$ probabilities are computed by another algorithm (Li and Reynolds, 1993; Riitters et al., 1996). Because the contagion metric is computationally intensive, for this exercise it would be tedious to determine this value by hand for even a relatively tiny landscape such as the early-settlement landscape. Thus, for illustration purposes, you will compute the contagion value for only a subset of that landscape.

CALCULATIONS

Metrics of Spatial Configuration in an Early-Settlement Landscape

Refer back to Figure 7.1. Recall that the early-settlement landscape is meant to represent an area that was formerly fully forested, although some of the land has been converted for agricultural and urban use.

Calculation 2.1: Using the four-neighbor rule, calculate the total number of patches for each cover type in the early-settlement landscape. Enter your results in Table 7.4 on the CD.

Calculation 2.2: Using the four-neighbor rule, calculate the mean patch size for each cover type in the early-settlement landscape. Enter your results in Table 7.4.

Calculation 2.3: Calculate the number of edges for each category in the early-settlement landscape of Figure 7.1. Be sure to count both horizontal and vertical edges between cover types; you may find it useful to mark edges in pencil in your lab manual as you count. Enter your results in Table 7.5 on the CD.

Calculation 2.4: Using the results from Calculation 2.3, compute the edge-to-area ratio for each cover type and enter into Table 7.5.

Calculation 2.5: To begin calculating contagion, use Figure 7.2 to calculate the proportions occupied by each of the three land cover types in the *subset* of the early-settlement landscape. Record the values in Table 7.6 on the CD.

Calculation 2.6: Count the adjacencies for all cover types for the *subset* of the early-settlement landscape, as seen in Figure 7.2. Enter the results in Table 7.7 on the CD. Do not count the borders of the map for this exercise. HINT: If you mark each adjacency once as it is counted, you will mark 40 adjacencies.

Calculation 2.7: Note that Table 7.7 is symmetric. In effect, we have counted many, but not all, of the adjacencies twice. In particular, diagonal elements, which represent adjacencies between cells of the same type, have been counted only once. So that each adjacency is counted the same number of times, double the values from the diagonal elements of Table 7.7

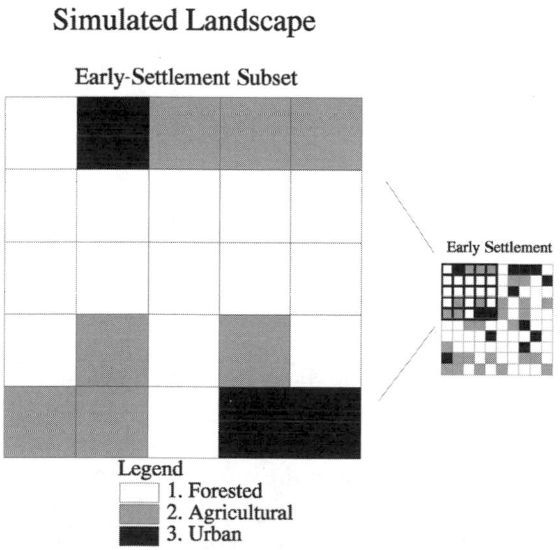

Simulated Landscape

Early-Settlement Subset

Legend
☐ 1. Forested
▨ 2. Agricultural
■ 3. Urban

FIGURE 7.2
Subset of the early-settlement landscape used for calculating the contagion index.

Early Settlement

and enter them in Table 7.8 on the CD, the **N** matrix. For the nondiagonal elements of Table 7.8, use the same value seen in Table 7.7.

Calculation 2.8: Use the values of the **N** matrix (Table 7.8) to compute the elements of the **Q** matrix (Table 7.9 on the CD).

Calculation 2.9: Calculate the contagion value for the subset of the early-settlement landscape using the elements of the **Q** matrix.

The contagion value: _____

Question 2.1. What characteristics of a landscape will influence the result you obtain for the number of patches and the average patch size?

Question 2.2. What characteristics of the landscape appear to have influenced the contagion value calculated in this section? How would you change the values of the grid cells to raise the contagion value?

Question 2.3. If you were considering a real landscape, do you think it would be reasonable, in general, to save computer time by calculating the contagion value for only a subset of the landscape? What characteristics of a real landscape might inhibit or encourage you to use only a subset?

Question 2.4. Suppose that you are given the task of describing how a landscape changed between two time periods, t_1 and t_2. The map of the first time period contains five cover types; the map from the second time period contains seven cover types because "forest" in t_2 was mapped in more detail—as deciduous, coniferous and mixed forest. How should you proceed with your comparison, and why?

Question 2.5. Two landscapes are the same size and both contain the same amount of a given cover type. Landscape A has 4 patches of that cover type, and landscape B has 17 patches of the same cover type. Which of the landscapes will have the greater length of edge of that cover type?

Question 2.6. From your set of calculations, do you think that after calculating a large number of metrics for a single landscape, additional metrics would provide little new information? How might you attempt to objectively determine an upper limit to the number of useful metrics?

Part 3. Using FRAGSTATS for Automated Landscape Metric Calculation for the Early and Post-Settlement Landscapes

In this section, you will use FRAGSTATS (McGarigal and Marks, 1995) to analyze the landscapes you examined in Parts 1 and 2. FRAGSTATS is available for free, computes a wide variety of metrics, is available in versions to analyze both raster and vector maps, and is probably the most widely used program for landscape pattern analysis. The FRAGSTATS program (Fragstat.exe) as well as the other files you will need for Parts 3 and 4 are on the CD-ROM under the directory "FRAGSTAT" for this lab. To avoid potential problems, we *strongly recommend* you copy the entire folder for the lab to your computer (not just individual files). Within this directory is another folder called "Document," which contains frag.pdf, the excellent documentation for the program. This document presents a conceptual overview of each metric. Also included are the appendices fraga.pdf, fragb.pdf, and fragc.pdf. (You are strongly encouraged to read these files, as FRAGSTATS will produce many more landscape metrics than you have computed by hand in this exercise.) You will input FRAGSTATS-compatible files (which have been given the arbitrary suffix .fra) for the early- and post-settlement landscapes. You will then be prompted for a variety of information from FRAGSTATS as detailed here to specify your output files and the algorithm to use for your calculations.

INPUT

(a) FRAGSTATS accepts ASCII files as input. We have specified landscape files with the .fra extension. FRAGSTATS will prompt:

```
Enter name of landscape image:
```

Select the appropriate .fra file for your FRAGSTATS run.

(b) FRAGSTATS then will prompt:

```
Enter base name for output files . . .
```

You may enter any string up to eight characters. Suggestions are given in each of the calculations for this section.

(c) FRAGSTATS will then lead you through a series of questions to specify settings for the suite of metrics it calculates for your image. The prompts and their meaning are given below.

```
Enter size of grid cell in meters {must be square}:
```

Enter the size of one side of a grid cell for your image. The size of cells for each image is given in each of the calculations for this section.

```
Enter distance from edge in meters to use in core area
determination:
```

In this section, we will not use core area metrics, so hit the Enter key to use the default value. See the documentation for core area metrics for further information.

```
Enter type of input image file:
```

Enter "2" to indicate an ASCII file.

```
Enter number of rows, columns in image:
```

Enter the appropriate number of rows and columns for the image (e.g., "10 10"). The number of rows and columns of each image is given in each of the calculations for this section.

```
Enter the value for interior (positive) background cells
(even if you only have exterior background):
Enter the maximum number of patch types in the landscape
(for relative patch richness):
Enter name of file containing weights for each combination
of patch type:
Enter option for patch ID image:
Enter name of ASCII file containing character descriptors
for each patch type:
What proportion (equivalent to contrast weight) of the
landscape boundary and background class edges should be
considered edge? . . .
```

Each of these questions is outside the scope of this section. Use the Enter key to enter the default value for each of these questions. The FRAGSTATS documentation has more information about each of these prompts if you want a deeper treatment of the capabilities of the program.

```
Use diagonals in patch finding [y/n]:
```

As directed in each calculation below, enter "y" to specify the eight-neighbor rule for finding patches, or "n" to specify the four-neighbor rule.

```
Enter search radius for proximity index:
```

In this section, we are not interested in proximity indices, so hit the Enter key to use the default value.

```
Calculate nearest neighbor distance:
```

We are not interested in knowing nearest neighbor distances, so hit the "n" key to use the default value.

```
Output patch level statistics [y/n]:
```

In this section, we are not interested in knowing details about each patch, but instead are primarily interested in metrics that summarize the entire image. Enter "n".

```
Output class level statistics [y/n]:
```

Although we will not directly use the information contained in the summaries of each landscape category, it is interesting to note that some metrics can be calculated for each class. Enter "y".

OUTPUT

FRAGSTATS outputs information in several files. In this lab, we are concerned with the .ful file, a text file that can be viewed with any text editor. Information about each landscape category is at the beginning of the file, and metrics for the entire landscape are at the end of the file. In these landscapes, category 1 = forested, category 2 = agricultural, and category 3 = urban.

CALCULATIONS

Calculation 3.1: The early-settlement landscape with the four-neighbor rule.

1. Run FRAGSTATS using the esett.fra landscape file. This is a 10 × 10 landscape in which one side of a cell represents 1000 meters on the ground.

2. As the base for output file names, enter "early4".

3. Compare the results given by FRAGSTATS with your answers calculated by hand for the early-settlement landscape in the previous section. (NOTE: FRAGSTATS does not calculate dominance.)

Calculation 3.2: The early-settlement landscape with the eight-neighbor rule.

1. Run FRAGSTATS using the eight-neighbor rule for the early-settlement landscape. Again, use the esett.fra landscape file.

2. As the base for naming output files, enter "early8".

3. Use the same options as in the previous FRAGSTATS run, except when asked about using diagonals for finding patches. At that prompt, enter "y" to invoke the eight-neighbor rule.

Calculation 3.3: The post-settlement landscape with the four-neighbor rule.

1. Run FRAGSTATS using the psett.fra landscape file. This is a 10 × 10 landscape in which one side of a cell represents 1000 meters on the ground.

2. As the base for output files, enter "post4".

3. Use the same options as in the previous FRAGSTATS runs.

Calculation 3.4: The post-settlement landscape with the eight-neighbor rule.

1. Run FRAGSTATS using the eight-neighbor rule for the post-settlement landscape. Again, use the psett.fra landscape file.

2. As the base for output files, enter "post8".

3. Use the same options as in the previous FRAGSTATS runs.

Question 3.1. Organize the results obtained for the four runs (early- and post-settlement landscapes, four- and eight-neighbor rules). Describe how the metrics are affected by the choice of the four- and eight-neighbor rule. Taken as a whole, how do the metrics indicate that this landscape has changed from the early-settlement to post-settlement period?

Part 4. Automated Landscape Metric Calculation for Real Landscapes

In this section, we use FRAGSTATS to compute landscape metrics for real landscapes provided on the CD-ROM. The provided landscapes differ in landscape size, number of classes, and spatial resolution. Two version of each image are provided. The FRAGSTATS-compatible files (.fra) are contained under the directory "FRAGSTATS" for this lab. The folder "IMAGES" contains color image files (*.pdf) that can be viewed and printed using Adobe Acrobat Reader (which can be downloaded for free from the Web).

CALCULATIONS

Calculation 4.1: Madison, Wisconsin
We present two classifications of the same satellite image produced by two different users of the same landscape processing software. Subjectivity inherent in the classification process inevitably produces differences among resultant maps. The two landscapes can be browsed using mad1.pdf and mad2.pdf. FRAGSTATS-compatible files are mad1.fra and mad2.fra. Each landscape has 575 rows and 800 columns, and one side of a grid cell represents 30 meters on the ground.

Calculation 4.2: Las Vegas and vicinity, Nevada.
This landscape can be browsed using lasveg.pdf, and the FRAGSTATS-compatible file is lasveg.fra. This landscape has 547 rows and 796 columns, and one side of a grid cell represents 100 meters on the ground.

Calculation 4.3: Superior National Forest and environs, Minnesota.
This landscape can be browsed using minne.pdf, and the FRAGSTATS-compatible file is minne.fra.

1. This landscape has 187 rows and 234 columns, and one side of a grid cell represents 1000 meters on the ground.
2. This landscape contains cells that are "background" to the classification; they represent Lake Superior to the southeast and Ontario, Canada, to the north. FRAGSTATS will prompt for the value to use for interior background cells:

```
Enter the value for interior (positive) background cells
(even if you only have exterior background):
```

At this prompt, enter "−9999". This will allow the program to recognize that cells outside the classification area, which have a value of −9999 in minne.fra, should be considered as background during metric calculations.

Calculation 4.4: Brazilian Amazon, South America.
This landscape can be browsed using braz.pdf, and the FRAGSTATS-compatible file is braz.fra. This landscape has 729 rows and 942 columns, and one side of a grid cell represents 1000 meters on the ground.

Question 4.1. Which of the landscapes in Part 4 shows the most fragmented pattern, and which is the least fragmented? How did you determine this?

Question 4.2. Using the results produced by FRAGSTATS for this set of landscapes, use a single graph to plot the values of each of the following landscape-level metrics:

Contagion

Patch density (the average number of patches per 100 ha)

Edge density (an expression of edge-to-area relationships)

Landscape shape index (a measure of shape complexity)

Largest patch index (an indicator of connectivity)

Patch richness (the number of patch types)

Note that you can use pencil and graph paper to plot these graphs, or enter the data into a spreadsheet program to produce the graphs and even compute correlation coefficients.

(a) By inspecting the plots, determine which landscape metrics appear to be correlated with each other.

(b) What metrics appear to be relatively independent?

(c) Why do you think these relationships occur?

Question 4.3. How would the correlation among landscape metrics influence your choice of what to report in an analysis that describes landscape pattern or quantifies differences between two landscapes or changes in a single landscape through time?

Question 4.4. What criteria would you use to select the "best" set of metrics to describe a landscape?

BIBLIOGRAPHY

Note. An asterisk preceding the entry indicates that it is a suggested reading.

BAKER, WILLIAM L., AND Y. CAI. 1992. The r.le programs for multi-scale analysis of landscape structure using the GRASS geographical information system. *Landscape Ecology* 7:291–302.

GARDNER, R. H., B. T. MILNE, M. G. TURNER, AND R. V. O'NEILL. 1987. Neutral models for the analysis of broad-scale landscape pattern. *Landscape Ecology* 1:19–28.

*GUSTAFSON, E. J. 1998. Quantifying landscape spatial pattern: What is the state of the art? *Ecosystems* 1:143–156. A synthetic overview of the ways in which landscape pattern is quantified, emphasizing conceptual issues and arguing for complementary use of patch-based measures and approaches from spatial statistics.

GUSTAFSON, E. J., AND G. R. PARKER. 1992. Relationships between landcover proportion and indices of landscape spatial pattern. *Landscape Ecology* 7:101–110.

*HAINES-YOUNG, R., AND M. CHOPPING. 1996. Quantifying landscape structure: A review of landscape indices and their application to forested landscapes. *Progress in Physical Geography* 20:418–445. A good review that includes examples of how different landscape metrics are used in questions associated with forested landscapes.

*HARGIS, C. D., J. A., BISSONNETTE, AND J. L. DAVID. 1998. The behavior of landscape metrics commonly used in the study of habitat fragmentation. *Landscape Ecology*

13:167–186. Hargis et al. examine whether commonly used landscape metrics can distinguish between landscapes that are fragmented and those that are not. This paper can provide a useful basis for a discussion of the behavior of landscape metrics.

LI, H., AND J. F. REYNOLDS. 1993. A new contagion index to quantify spatial patterns of landscapes. *Landscape Ecology* 8:155–162.

LI, H., AND J. F. REYNOLDS. 1994. A simulation experiment to quantify spatial heterogeneity in categorical maps. *Ecology* 75:2446–2455.

*LI, H., AND J. F. REYNOLDS. 1995. On definition and quantification of heterogeneity. *Oikos* 73:280–284. An excellent discussion of what is meant by heterogeneity; it should be read by all those beginning to consider the causes or consequences of spatial pattern.

MCGARIGAL, K., AND B. J. MARKS. 1995. Fragstats: Spatial analysis program for quantifying landscape structure. USDA Forest Service General Technical Report PNW-GTR-351. USDA Forest Service, Pacific Northwest Research Station, Portland, OR.

O'NEILL, R. V., J. R. KRUMMEL, R. H. GARDNER, G. SUGIHARA, B. JACKSON, D. L. DEANGELIS, B. T. MILNE, M. G. TURNER, B. ZYGMUNT, S. CHRISTENSEN, V. H. DALE, AND R. L GRAHAM. 1988. Indices of landscape pattern. *Landscape Ecology* 1: 153–162.

PIELOU, E. C. 1975. *Ecological Diversity*. Wiley-Interscience, New York.

*RIITTERS, K. H., R. V. O'NEILL, C. T. HUNSAKER, J. D. WICKHAM, D. H. YANKEE, S. P. TIMMONS, K. B. JONES, AND B. L. JACKSON. 1995. A factor analysis of landscape pattern and structure metrics. *Landscape Ecology* 10:23–40. Provides the first comprehensive analysis of a wide range of landscape metrics, demonstrating how strongly many of the metrics are correlated and identifying five independent aspects of pattern.

RIITTERS, K. H., R. V. O'NEILL, J. D. WICKHAM, AND K. B. JONES. 1996. A note on contagion indices for landscape analysis. *Landscape Ecology* 11:197–202.

ROMME, W. H. 1982. Fire and landscape diversity in subalpine forests of Yellowstone National Park. *Ecological Monographs* 52:199–221.

ROMME, W. H., AND D. H. KNIGHT. 1982. Landscape diversity: The concept applied to Yellowstone Park. *BioScience* 32:664–670.

TURNER, M. G., R. COSTANZA, AND F. H. SKLAR. 1989. Methods to compare spatial patterns for landscape modeling and analysis. *Ecological Modelling* 48:1–18.

*TURNER, M. G., R. H. GARDNER, AND R. V. O'NEILL. *Pattern and Process: Landscape Ecology in Theory and Practice*. Springer-Verlag, New York, chapter 4.

WICKHAM, J. D., AND D. J. NORTON. 1994. Mapping and analyzing landscape patterns. *Landscape Ecology* 9:7–23.

*WICKHAM, J. D., AND K. H. RIITTERS. 1995. Sensitivity of landscape metrics to pixel size. *International Journal of Remote Sensing* 16:3585–3594.

CHAPTER 8

Understanding Landscape Metrics II

Effects of Changes in Scale

Joshua D. Greenberg, Sarah E. Gergel,
and Monica G. Turner

OBJECTIVES

The perceived pattern of a landscape is influenced by the scale at which the landscape is represented, either in an aerial photograph, satellite image, or data in a Geographic Information System (GIS). This effect of scale has serious implications for the analysis and interpretation of landscape metrics. As such, understanding the range of variability in landscape metrics solely due to changes in scale, and why some metrics are more sensitive to these changes than others, is critical to the appropriate use of landscape metrics. The primary goals of this lab are to

1. explore the quantitative effects of changes in scale (grain and extent) and classification scheme on the computation of landscape metrics; and
2. examine the implications of metric scale dependence for the interpretation of landscape metrics.

In Part 1, to ensure that you fully understand the mechanics behind the different changes in scale, you will change the grain, extent, and classification scheme by hand on a small hypothetical landscape. It is recommended that you work in pairs for this section to divide the workload. In Part 2 you will use the ESRI program ArcExplorer, to examine the effects of changes in scale and classification scheme on landscape metrics by using a large landscape image. In the latter section, the changes in scale and the landscape metric calculations (in FRAGSTATS) have already been performed for you. Your responsibility will be to analyze and interpret the data resulting from the scale changes. Part 1 of this lab requires pencil, paper, calculator, overhead transparencies, and colored markers. Part 2 of this lab requires a PC running ArcExplorer, a spreadsheet program, the Excel file **scale.xls**, and the ArcExplorer files accompanying the laboratory located on the CD. *It is essential that you copy the entire, intact subdirectory for this chapter, including **ArcData**, to your own PC in order to retain all the necessary files for Part 2.* Familiarity with Chapter 7, Understanding Landscape Metrics I, is required, and familiarity with ArcExplorer as in Chapter 3, Introduction to GIS, or ArcView software is extremely helpful.

INTRODUCTION

Quantification of spatial pattern in landscapes is necessary to evaluate the differences among landscapes and track changes through time. Many factors can influence the results of a pattern analysis, but two extremely important factors are the scale of the data—both its grain and extent—and the way in which the landscape was classified. **Scale** refers to the spatial or temporal dimensions of an object or process, and it is characterized by both grain and extent. Spatial **grain** is the finest level of spatial resolution possible within a given data set. For example, grain refers to the cell size for gridded maps, or the minimum polygon size for vector maps. Spatial **extent** refers to the size of the overall study area. **Classification scheme** refers to the number and type of classes used in grouping the landscape cover types. For example, a landscape might be classified into forested and nonforested categories, or be separated further into forested, urban, agricultural, and prairie categories.

The grain and extent of the data used in any analysis of landscape pattern influence the numerical result obtained for a given metric (Turner et al., 1989; Moody and Woodcock, 1995; Wickham and Riitters, 1995; O'Neill et al., 1996). This sensitivity means that comparison of landscape data represented at different scales may be invalid, or might primarily reflect scale-related differences rather than actual differences in landscape pattern. Moody and Woodcock (1995) changed the grain size of a Landsat image and demonstrated that scale-induced changes in the proportion of the landscape occupied by different cover types influenced landscape metrics such as patch size, patch density, and landscape diversity. However, not all metrics are equally sensitive to changes in grain (Cain et al., 1997). Furthermore, spatial extent can influence landscape metrics independently of grain size. Smaller maps may show the effects of map boundaries, resulting in biased measurements of patch size and shape. As with the long-recognized relationship between the number of species observed in a given location as a function of area (i.e., species-area curves), landscapes of larger extent often contain a greater number of cover types.

Finally, the classification scheme used when a landscape is represented as a map also strongly influences the numerical results of any pattern analysis (Wickham et al., 1997). The categories should be selected for a particular question or purpose. For example, twelve general categories of land cover might be appropriate to study landscape patterns in the Upper Midwest or New England regions of the United States, but the spatial heterogeneity within smaller areas, such as a state park, would require more finely distinguished vegetation classes. In this lab, we will illustrate how the same landscape can look quite different under alternative classification schemes.

EXERCISES

Part 1. Understanding Changes in Scale

To ensure that you understand how the maps you will analyze later were produced, you will first conduct a series of scale changes by hand on a small landscape image. The landscape will be created by you, using Figure 8.1 on the CD. This base map represents a 12 × 12 pixel area and consists of four cover types: (1) forest, (2) wetland, (3) water, and (4) roads. Print a hard copy of Figure 8.1 and follow the simple instructions to create your base map.

EXERCISE 1.1
Changes in Grain

The grain size of spatial data has a profound influence on how a landscape is quantified with metrics. While a finer grain provides more detail, this level of detail must be balanced by the sheer quantity of the resulting data relative to disk storage space.

You will now create two new maps from your original base map, with each successive map at a coarser grain.

1. Use the completed version of Figure 8.1 as (12 × 12 pixels with 144 total pixels, grain = 30 m on a side of each pixel).

2. Aggregate the base map to create a new 6 × 6-pixel map (36 pixels total, grain = 60 m). The total extent of the region stays the same, but fewer pixels are used to represent the area.

3. Assign a cover type category to each new pixel in the new 6 × 6-pixel map based on a majority rule. That is, if the new pixel contains three pixels of forest and one pixel of urban from the base map, the pixel in the new map would be classified as forest. If there is a tie, flip a coin.

4. Repeat this process, but aggregate the 6 × 6-pixel map to a 3 × 3-pixel map (grain = 120 m).

5. Calculate the proportion (p_i) of each cover type in each of the three landscapes. Record your results in Table 8.1, which can be printed from the CD.

6. Calculate mean p_i value for each cover type (across all landscapes) and enter in Table 8.1.

7. Calculate Coefficient of Variation (expressed as a percent; CV%) for each cover type across all landscapes. Enter your results in Table 8.1.

Question 1.1. What happened to the p_i of the rarest and most dominant cover types?

Consider the aggregation rule used in this exercise as you complete the rest of the lab. Think about other possible aggregation rules and when they might be useful.

EXERCISE 1.2
Changes in Extent

Now change the extent of your base map to successively smaller map sizes while keeping the grain size constant. A decision will have to be made about how to reduce the extent of your image—either starting at one corner or reducing the area centered on the middle of the image.

1. Start with Figure 8.1 as your base map (extent = 360 m on a side of image).
2. Reduce the extent of the image to 6 × 6-pixel area (extent = 180 m on a side, with each pixel 30 m in size). Remember that the grain remains the same, and you do not alter any cover type assignments.
3. Reduce the extent of the image to a 3 × 3-pixel area (extent = 90 m).
4. Calculate the proportion (p_i) of each cover type for each landscape extent. Record your results in Table 8.2 on the CD.
5. Calculate mean p_i values for each cover type over all extents and enter in Table 8.2.
6. Calculate CV% for each cover type over all extents and enter in Table 8.2.

Question 1.2. What happened to the total number of cover types in the landscape?

As you work through the rest of the exercises, consider the cover types present in the landscapes of different extent. Think about the possible implications of using maps of different extent.

EXERCISE 1.3
Changes in Classification Scheme

Next, you will enact changes in classification by creating new cover types and/or merging existing cover types.

1. Start with Figure 8.1 (on the CD) as your base map (four cover types). Do not alter the grain or extent of the map.
2. Reclassify this four-cover-type map to a map with three categories. Carefully consider your reasoning for your reclassification.
3. Now reclassify the base map with four cover types to a binary (two-category) map.

Question 1.3. How did you decide which categories to use when reclassifying the map? Was this an arbitrary decision? Why or why not?

Part 2. Effects of Changes in Scale on Landscape Metrics

In this section of the lab, you will examine the effects of changes in scale and classification scheme on a suite of landscape metrics calculated for spatial data from an actual landscape. Much of the laborious work and data collection has already been completed for you; your goal is a synthetic interpretation of the results. The changes in scale and classification scheme that have been performed on the spatial data are virtually identical to the changes you enacted in Part 1 and can be viewed using ArcExplorer. A subset of landscape metrics were calculated for each landscape using FRAGSTATS, and the results were compiled in a spreadsheet file, **scale.xls.** You will be responsible for observing the spatial data (and accompanying scale changes) using ArcExplorer, analyzing and graphing the spreadsheet metric data, and answering a series of question regarding the influence of changes in scale and classification scheme on landscape metrics.

You will analyze the spatial data used in Chapter 3, Introduction to GIS, representing a 25-km² area of the Gifford Pinchot National Forest in southwest Washington state (USA). Recall that the Yacolt burn (an extensive fire in 1903) left a mosaic of burned and unburned patches in the area. Patches not burned by the fire are now considered old-growth forests (>200 years old), while burned areas are considered either mature (Age 80–200 years old) or young (Age 41–80 years old) forest. Forest patches less than 80 years old are likely the result of timber harvest. See the metadata file (**metadata.pdf**) included on the disk under the directory of Chapter 3 for a more complete description of the themes. *(NOTE:* In addition to the Excel spreadsheet **scale.xls** and the directory **ArcData** needed for this lab, we have also provided the folder **FragData** on the CD, which provides access to the FRAGSTATS grids and output files on which this lab is based.)

EXERCISE 2.1
Effects of Changes in Grain

1. Open ArcExplorer by double-clicking on its icon. (See the Chapter 3, Introduction to GIS, for directions on installing and using Arc/Explorer.)

2. Use **File** then **Open Project** on the pull-down menu to open the project file **Grain.AEP** from the ArcData subdirectory. This project contains themes showing grain changes in the spatial data with theme names that correspond to the size of the grain. For example, the Veg50 theme has a grain size of 50 meters (i.e., one side of a pixel is 50 m). The method used to change the grain of these themes was a majority method, which takes the dominant cover class in the finest-scale theme and gives that class to the new, larger cell.

3. Familiarize yourself with the different themes and visually examine the changes in pattern with changes in grain size. When viewing several themes, remember that ArcExplorer displays the themes starting with the ones listed at the bottom of the legend and placing subsequent themes on top. In some cases it may be necessary to uncheck the display box for a theme higher on the legend to see the themes listed lower on the legend. It may be helpful to print these maps if you have access to a color printer.

4. Using a spreadsheet program, open the file **scale.xls**.

5. Be sure you are on the sheet titled **Grain-Summary.** If not, click on the tabs on the bottom of the spreadsheet.

6. Examine the data for dominance in the spreadsheet, which is also shown in the plot in Figure 8.2. Answer Questions 2.1 and 2.2 below.

7. Examine Figure 8.3 as well as the data on which it is based (in the spreadsheet). Answer Quesion 2.3 below.

8. Use the Grain-Summary data sheet to create the following graphs of landscape metrics for classes Age41–80 and AgeOver200 (note that a few additional metrics are also included in the spreadsheet).
   (a) Proportion (p_i) (on the y-axis) vs. grain (on the x-axis)
   (b) Total edge vs. grain
   (c) Mean patch size vs. grain

9. Calculate the CV% for each metric in step 8 for each cover type across all grain changes.

10. Enter your results in Table 8.3 from the CD and answer Question 2.4 below.

Question 2.1. Explain the trend in dominance with increasing grain size. Why does this occur?

Question 2.2. Note the dominance measure calculated for the landscape with a grain size of 200 meters. What do you see? Can you offer an explanation?

Question 2.3. Describe and interpret the trend in contagion with increasing grain size.

FIGURE 8.2
Changes in the grain (cell size) of an example map and the resulting influence on the dominance value.

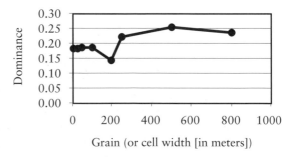

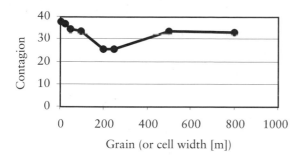

FIGURE 8.3
Changes in the grain (cell size) of an example map and the resulting influence on the contagion value of the landscape.

Question 2.4. For each of the graphs in step 8, describe how the metrics change with grain. Which metrics change the most with increasing grain? Which change the least?

Question 2.5. Which land-cover types are the least affected by changes in grain? Why is this?

Question 2.6. Suggest some principles for how landscapes appear to change when measured with increasing grain sizes.

EXERCISE 2.2
Effects of Changes in Extent

1. Open the file **extent.AEP** using ArcExplorer. In this project, the extent of each theme (measured here as the width of the entire study area) has been changed, centered on the middle of the theme. Use the measuring tool to determine the approximate extent of each theme—see Chapter 3 for a reminder on how. Recall the software bug from Chapter 3—do *not* measure in kilometers. Note how changing the extent changes the visual appearance of the data.

2. Examine the dominance and contagion metrics in the **Extent-Summary** data page in the spreadsheet.

3. Last, examine the plots of dominance versus extent (Figure 8.4) and contagion versus extent (Figure 8.5). Answer Question 2.7 below.

4. Use the **Extent – Summary** sheet to create the following graphs of landscape metrics for classes Age41–80 and AgeOver200.
   (a) Proportion (p_i) (on the y-axis) vs. extent (on the x-axis)
   (b) Total edge vs. extent
   (c) Mean patch size vs. extent

5. Calculate the CV% for each metric in step 4 for both cover types across all extent changes.

6. Enter your results in Table 8.4 and answer Question 2.8 below.

F<small>IGURE</small> 8.4
Changes in the extent of an example map and the impact on the dominance value.

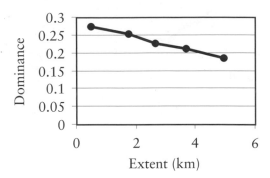

F<small>IGURE</small> 8.5
Changes in the extent (or total area) of an example map and the impact on the contagion value of the landscape.

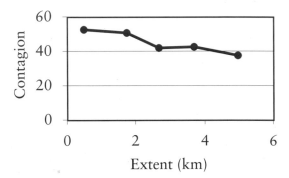

F<small>IGURE</small> 8.6
The influence of changes in classification scheme (from eight to five to three classes) on the dominance value of the landscape.

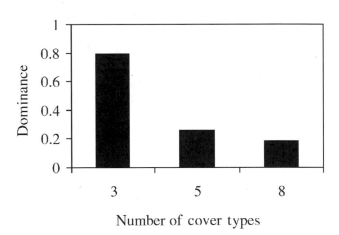

Question 2.7. Examine the value of dominance for the smallest extent. What do you see and why is this so? Does contagion vary with extent? Interpret the trend that you observe.

Question 2.8. For each of the graphs in step 6, describe how the measures change with extent. Which metrics change the most with increasing extent? Which change the least?

Question 2.9. Which land-cover types are the least affected by changes in extent? Why is this?

Question 2.10. Explain how changes in extent alter the patches on the edge of the extent, and what could be done to help remedy that effect in pattern analysis.

EXERCISE 2.3
Changes in Classification Scheme

The number of classes used in a GIS theme (or any map) is often a subjective decision made by the map creator. In this exercise, you will examine changes in landscape patterns due to changes in the classification scheme of the vegetation.

1. Open the file **class.AEP**. In each reclassification, the bare ground and hardwood classes have been maintained, while the six age classes of the evergreens have been reduced from six classes to three classes to one class. The theme names, **veg3class**, **veg5class**, and **veg8class**, refer to the total of number of classes used in the classification. Answer Question 2.11 below.

2. Examine the Class-Summary sheet for both dominance and contagion.

3. Compare these data to the plots in Figures 8.6 and 8.7.

Question 2.11. Visually, give a qualitative estimate of how this landscape changes with the different classification schemes.

Question 2.12. Explain the changes in dominance and contagion when the number of classes are changed.

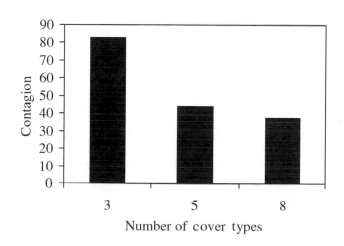

FIGURE 8.7
The influence of changes in classification scheme (from eight to five to three classes) on the contagion value of the landscape.

Question 2.13. Will dominance and contagion always change in the same way as classes change, or will variations in the metrics be specific to the landscape?

CONCLUSIONS

Quantitative measures of landscape pattern are useful for a variety of purposes. However, when analyzing or interpreting landscape metrics, one must pay close attention to the scale of the data and the classification scheme used. Differences in scale among two data sets (e.g., between time periods or for different landscapes) can obscure or augment the actual amount of landscape change and limit the comparability of many landscape metrics. Research into the behavior of metrics must identify the relative sensitivity of various metrics to scale differences and work toward developing scaling rules for extrapolating across different scales.

DISCUSSION QUESTIONS

1. The aggregation rule used to change grain size for this lab was a majority rule. What are two other rules that could have been used? Explain the advantages and disadvantages of each and the conditions under which one might choose to apply each.
2. Compare the CV% of dominance across grain, extent, and classification scheme changes.
   (a) Which alteration resulted in the most variance in dominance? Why?
   (b) Which alteration resulted in the least variance in dominance? Why?
3. Compare mean patch size for each landscape (regardless of cover type). Does grain, extent, or classification scheme have the greatest influence on mean patch size? Why?
4. Compare the relative impact of changes in grain versus changes in extent for the metrics you examined (refer to your answers for Questions 2.4 and 2.8). Are any metrics insensitive to changes in both grain and extent? What attributes of a metric might reduce its sensitivity?
5. The amount of edge habitat in a landscape is often quantified for wildlife-related research because many species respond very differently to edge habitats. If quantifying edge habitat for white-tailed deer, what considerations would you suggest regarding selection of grain, extent, and number of classes for the study landscape?
6. Should spatial pattern analyses be conducted across multiple scales? What is the utility of conducting a spatial pattern analysis at only one scale? What is the utility of conducting a pattern analysis at more than one scale?

BIBLIOGRAPHY

Note. An asterisk preceding the entry indicates that it is a suggested reading.

CAIN, D. H., K. RIITTERS, AND K. ORVIS. 1997. A multi-scale analysis of landscape statistics. *Landscape Ecology* 12:199–212.

*MEENTEMEYER, V., AND E. O. BOX. 1987. Scale effects in landscape studies. In M. G. Turner, ed. *Landscape Heterogeneity and Disturbance.* Springer-Verlag, New York, pp. 15–34. One of the earlier papers addressing scale effects in landscape studies; it lays out a variety of ways in which scale is important.

MOODY, A., AND C. E. WOODCOCK. 1995. The influence of scale and the spatial characteristics of landscapes on land-cover mapping using remote sensing. *Landscape Ecology* 10:363–379.

O'NEILL, R. V., C. T. HUNSAKER, S. P. TIMMINS, B. L. JACKSON, K. B. JONES, K. H. RITTERS, AND J. D. WICKHAM. 1996. Scale problems in reporting landscape pattern at the regional scale. *Landscape Ecology* 11:169–180.

TURNER, M. G., R. V. O'NEILL, R. H. GARDNER, AND B. T. MILNE. 1989. Effects of changing spatial scale on the analysis of landscape pattern. *Landscape Ecology* 3: 153–162.

*WICKHAM, J. D., R. V. O'NEILL, K. H. RITTERS, T. G. WADE, AND K. B. JONES. 1997. Sensitivity of selected landscape pattern metrics to land-cover misclassification and differences in land-cover composition. *Photogrammetric Engineering and Remote Sensing* 63:397–402. A useful investigation of the effects of classification scheme on resulting landscape metrics.

*WICKHAM, J. D., AND K. H. RITTERS. 1995. Sensitivity of landscape metrics to pixel size. *International Journal of Remote Sensing* 16:3585–3594. This paper empirically demonstrates the sensitivity of different landscape metrics to changes in grain, which is relevant to using satellite imagery.

CHAPTER 9

NEUTRAL LANDSCAPE MODELS

ROBERT H. GARDNER AND STEVEN WALTERS

OBJECTIVES

Spatial patterns in landscapes are influenced by interactions between biotic and abiotic processes, and vice versa. Exploration of the relationships between pattern and process in landscapes is often best accomplished using models. Neutral landscape models—models lacking consideration of ecological effects on landscape pattern (Gardner et al., 1987)—are helpful in characterizing how and to what degree ecological processes affect observed landscape patterns. Neutral models serve as a null hypothesis, or baseline, for comparison with actual landscapes. This lab is specifically designed to

1. illustrate the methods used for generating and analyzing patch structure in neutral landscape models;
2. demonstrate the effects of using different neighborhood rules when identifying landscape patches;
3. explore the factors influencing connectivity in landscapes, and explore threshold effects in connectivity; and
4. examine the use of neutral models for formulating hypotheses regarding the relationship between pattern and process in actual landscapes.

The analyses and comparisons of landscape patterns (both real and artificial) require a method for characterizing pattern (i.e., determining what constitutes a patch) and a set of quantitative measures of pattern. In this lab, the concept of connectivity in landscapes will be addressed, particularly with respect to the neighborhood rules used to define "patches," or "clusters," in a landscape. A variety of neutral models will be generated to create a broad range of landscape patterns. You will also become familiar with a number of common metrics used to quantify patterns in these landscapes. To accomplish these objectives, you will be using RULE, a software package for the generation of neutral landscape models and analysis of landscape pattern. The following is required for this lab: a computer running MS-Windows operating system (a Pentium with at least 32MB of RAM is recommended), a copy of RULE software, spreadsheet software (examples given use Excel), and familiarity running DOS software and commands within a Windows environment.

INTRODUCTION

Neutral, or null, models in ecology provide a useful baseline for comparison when examining potential cause-and-effect relationships. In terms of landscape pattern, a **neutral model** is one that exhibits characteristic spatial patterns in the absence of processes that may affect patterns in real landscapes (e.g., topography, resource gradients, and disturbance regimes) (Gardner et al., 1987; With and King, 1997). In the neutral models examined here, landscape pattern occurs either as simple random processes (random maps), or as a result of spatially correlated processes using algorithms derived from fractal geometry (multifractal maps). Thus, comparisons of patterns and landscape indices for real landscapes with those of neutral models can provide insight into the effect of the influence of ecological processes on landscape patterns; if analyses of a real landscape are significantly different from the appropriate neutral models, it is quite likely that some important ecological process is driving the patterns observed in the real landscape.

The analysis of landscape pattern is often accomplished by converting continuous land-cover data into a gridded map for analysis of pattern by computer programs. Most analysis methods involve the identification of habitat patches (or clusters) and the description of their sizes, shapes, and spatial arrangements. Although the clustering of habitat into patches may be visually apparent, clear rules must be established to allow computers to identify habitat patches uniquely and unambiguously.

The usual rule for patch definition is referred to as the "nearest-neighbor rule" (Figure 9.1a). The **nearest-neighbor rule** states that if two similar sites have one edge along the four cardinal directions in common, then they are members of the same patch. Iterative application of the rule to each site will result in the identification of all neighbors (and neighbors of neighbors) and, thus, the size and property of the patch. Because this rule assumes that sites must touch along at least one edge to be members of the same patch, a row of sites arranged along a direction diagonal to the grid axes (i.e., a long, thin patch stretching diagonally across the map) will not be identified as a single continuous patch. If these sites were to be rotated 45°, then they would be

FIGURE 9.1
Three major neighbor-
hood rules: (a) nearest
neighbor, (b) next-near-
est neighbor, and (c)
third-nearest neighbor.

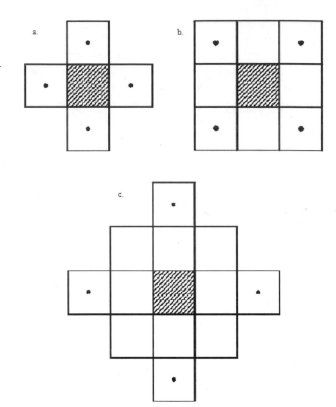

aligned along adjacent edges and identified as a single, continuous, connected site.

This problem can be corrected by using other rules for patch definition. The **eight-neighbor rule** for patch identification (Figure 9.1b) states that similar habitat sites are members of the same cluster if they touch along either the cardinal or diagonal directions. The **third-nearest-neighbor rule** (Figure 9.1c) extends consideration to sites that may not directly touch!

The patch identification process (with a given rule) is repeated to identify all other patches in a landscape. Measures of map pattern are provided by the analysis and summary of the statistical characteristics of all the patches. We will be performing exercises here that use a program, RULE, that creates random maps and allows alternative definitions of neighborhood rules to be applied to the analysis of landscape patterns.

Description of RULE

RULE is a DOS-based program originally written to investigate the properties of random maps, which serve as neutral models for the analysis of landscape pattern (Gardner et al., 1987; Gardner, 1999). The current version of RULE is written in Fortran-90 and runs under the MS-Windows environment. Table 9.1 illustrates a typical interactive dialog with RULE. Habitat patches are identified and analyzed within RULE depending on a user-specified neighborhood rule (hence, the program name), and the results are statistically summarized. The current version of RULE allows the user to specify either nearest-, next-nearest-, or third-nearest-neighbor rules, or to input relative coordinates for more complex rules. For each map generated and analyzed, a series of statistical summaries are produced (Table 9.2). For complete documentation on RULE, see Gardner (1999).

Generation of Random Maps

Several types of maps can be generated by RULE. The ones of interest for this exercise are simple random maps and multifractal maps (Figure 9.2). Maps created by other programs, or developed from GIS layers, may also be read into RULE for analysis of spatial patterns.

Simple random maps (Figure 9.2a) are created by specifying the number of rows and columns in the map, the number of habitat types to be generated, and the probabilities associated with each habitat type—including the probability for areas lacking any habitat at all. Examine Table 9.1 for a sample dialog producing a simple random map. The process of map generation is, as the name implies, a simple one. A uniform random number (i.e., a computer-generated number ranging from 0.0 to 1.0) is used to randomly assign a habitat type to each grid site. For instance, if the user specifies that p, the probability that an individual site will be equal to a specific habitat type i, is equal to 0.5, then a random number between 0.0 and 1.0 is generated for each grid site. If this random number is ≤ 0.5, then the habitat type of that site is set to i.

Multifractal maps are the most realistic landscapes generated by RULE (Figures 9.2b, 9.2c, and 9.2d). These maps seem so realistic because they are produced by a fractal algorithm that generates spatially correlated patterns of land cover. These maps are most often selected by investigators wishing to use random, but realistic, maps to simulate the movement and dispersal of organisms (With et al., 1997). Multifractal maps (Figures 9.2b, 9.2c, 9.2d) are generated in RULE by the midpoint displacement algorithm (MidPointFM2D, Saupe, 1988:101). This algorithm creates a map of real numbers by iterative interpolation to locate the midpoint of a line, perturbation of the midpoint by a Gaussian random value (GRV), and successive reductions of the variance of the GRV. Two parameters are used by RULE to control the generation of multifractal maps:

1. L, the "number of levels" or iterations of the midpoint displacement algorithm. Thus, the size of the map is always equal to 2^L. For instance, when $L = 4$, then the dimensions of the map (number of rows and columns) = 16; when $L = 6$, map dimensions = 64, etc.

2. H, the parameter that controls the rate of reduction of the GRV in successive iteration of the midpoint displacement method.

TABLE 9.1
SAMPLE RULE DIALOG PRODUCING A SIMPLE RANDOM MAP. USER
RESPONSE TO QUESTIONS BY RULE ARE GIVEN IN BOLD ITALIC. ALL
OUTPUT BY RULE IS WRITTEN TO A DISK FILE, RULERUN.LOG.

Neighborhood Analysis

Enter map type to be analyzed:

  <I> Input existing map file

  <R> Generate a random map (with replacement)

  <S> Generate a simple random map

  <C> Generate a curdled (hierarchical) map

  <M> Generate a multifractal random map

  <G> Generate a multifractal random map with a gradient

s

  Map choice: s

  Enter the number of map rows and columns

128 128

  Rows × Columns 5 128 × 128

  Enter a negative random number seed

−456587

  Random number seed: −456578

  Enter the neighborhood rule

    1—nearest neighbor (N_nb = 4)

    2—next nearest neighbor (N_nb = 8)

    3—3rd nearest neighbor (N_nb = 12)

    4—user defined

1

  Rule choice is: 1

  Enter the number of map classes

1

  Map classes =   1

  Type in the 0th probability.

0.5

  0.500000

  Type in the 1th probability.

0.5

  0.500000

    p(0) =   0.5000

    p(1) =   1.0000

  Enter the number of replications

10

```
N_Reps  =    10
Create an output maps?
  N = None
  G = generated map
  S = cluster Size map
  I = cluster ID map
  B = both
```

n

```
Map output choice = N
Perform map analysis?
<L>acunarity, <R>ule, or <A>ll?
```

r

```
Analysis method: R
Visualize map simulations <Y,N>
```

n

The generation of successive Gaussian increments results in a variance between points separated by distance x that is approximately equal to x^{2H} (assuming that σ^2, the variance of the Gaussian process, is equal to 1.0). Thus, values of H that range between 0.0 and 1.0 will result in maps that range from extremely fragmented (differences are negatively correlated, Figure 9.2b) to highly aggregated (differences are positively correlated, Figure 9.2d). For further details regarding the use of multifractal maps, see Plotnick and Gardner (1993), Pearson and Gardner (1997), and With et al. (1997).

Instructions for RULE

RULE and accompanying files can be found on the enclosed CD under the directory for Chapter 9.

RULE.exe is the RULE executable file that creates and analyzes maps with rows and columns of 514 or less (264,196 grid sites).

random.scr is an example script file to create and analyze a single random map with 128 rows and 128 columns.

fractal.scr is an example script file to create and analyze ten multifractal maps each with 128 rows and columns.

quick.bat is a sample batch file to sequentially run RULE for the analysis of different maps.

patgrid.map is a gridded map of forest patches within the Patuxent River watershed.

TABLE 9.2
INDICES FOR SPATIAL ANALYSIS OF MAP PATTERNS PRODUCED BY RULE
FOR EACH HABITAT TYPE. SEE GARDNER (1999) FOR ADDITIONAL
DETAILS CONCERNING THE CALCULATION OF EACH INDEX.

Index	Definition
Largest cluster size	Size of largest cluster (total number of grid units making up the largestcluster)
Largest cluster edge	Number of edges of largest cluster sites adjacent to a different habitat type
Largest cluster fractal	Fractal index of largest cluster estimated as ln(L.C.edge) / ln(average diameter of the cluster)
Largest cluster_rms	Mean squared radius of largest cluster (also known as the radius of gyration, Stauffer and Aharony, 1992). If r_i is the ith of s sites in the cluster, then L.C._rms = $\Sigma(r_i - r_j)^2 / s^2$. Diffuse sites of size s will have a larger L.C._rms than more compact sites.
Total clusters	Total number of clusters on the map
Total edges	Total edge of all clusters
Sav size	Area weighted average cluster size. If S_i is the size of the ith cluster, then Sav = $\Sigma S_i^2 / S$.
S_Freq	Total number of sites of current habitat type. P, the fraction of sites of the current habitat type, are estimated as $P =$ S_Freq / (nr * nc), where nr and nc are the number of rows and number of columns of the map, respectively.
Cor_len	Average mean squared radii of all clusters
Perc/freq	Frequency (percent of all maps) with a cluster large enough to span the dimensions of the map

Running RULE requires some familiarity with MS-DOS. Assuming you are running Windows software, open a DOS window. Create a directory for running RULE on your C: drive and copy the files from the CD to that directory. RULE is then run by simply typing **"RULE"** and answering the questions that the program asks.

NOTE: If you need to abort a run of RULE before it is finished, always use the **Control-C** *not* **Control-Alt-Delete** to end the session. Otherwise, you may have to recopy RULE from the CD to a new folder for RULE to operate properly.

INPUT

The generation of maps, and subsequent analysis, requires that RULE be provided with the following information:

Type and size of map to be generated

Initial value for the pseudorandom number generator

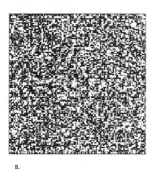
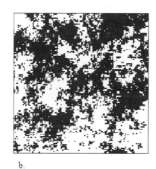

a.
b.

FIGURE 9.2
Sample maps produced
in RULE: (a) a simple
random map, and multi-
fractal maps with
(b) $H = 0.1$,
(c) $H = 0.5$, and
(d) $H = 0.9$.
In each instance, the
value of p, the propor-
tion of black cells within
the map, is equal to 0.5.

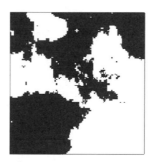

c.
d.

Neighborhood rule to use for map analysis

Number of habitat types and the probabilities of occurrence associated with each habitat type

Number of maps to be generated and analyzed

Type of analysis and output desired

Table 9.1 shows a sample interactive dialog to produce a random map with 128 columns and rows. The user may chose to view the created maps and patch patterns by answering "Y" to the final question asked by RULE (Table 9.1).

The script file, **random.scr,** is provided to illustrate running RULE in batch mode. To run in batch mode simply type "RULE **< random.scr**" You can use a text editor to change the specifications in the script file, produce multiple iterations of the same map type, analyze maps with different neighborhood rules, and so on. In addition, several script files may be created and sequential runs of RULE performed with a batch file. An example batch file, **quick.bat,** is provided for those familiar with this feature of DOS.

OUTPUT

The output for RULE is written to the screen and also saved to three data files: rulerun.log, cfd_scr.dat, and rustats.dat.

The main file of interest in this lab is **rulerun.log.** This file contains a list of the parameters input into RULE by the user (or the script file, if one is used), as well as a statistical summary of the maps analyzed in RULE. Table 9.2 lists the landscape metrics provided in rulerun.log. For each of these metrics, rulerun.log lists the mean, standard deviation, coefficient of variation, and minimum and maximum values. Note that if only one map is analyzed (either as an input map or a map produced from a single iteration in RULE), the mean, minimum, and maximum are all equal; the standard deviation and the coefficient of variation will be 0. In addition, if multiple habitat types are included, RULE produces within a single rulerun.log file a separate table of indices for each habitat type. *The file is overwritten every time RULE is executed;* thus, the user must rename and save the file for later use.

The file **cfd_scr.dat** is used to generate a cumulative frequency distribution (cfd) for landscape data. The file contains summary data for every patch identified and analyzed in a single execution of RULE; these data include the cluster number, its size, number of edges, fractal dimension, and radius mean squared (RMS). It includes a list of data for each individual map, and then lists of data divided by habitat type for each map. Because the file contains data for every patch, it can easily become a very large file for large or complex maps. *This file is overwritten with every execution of RULE,* and thus must be renamed to be saved.

The file **Rustats.dat** contains the data used to calculate the summary statistics found in rulerun.log, formatted to be easily imported into a spreadsheet or graphing software package. In order, the columns represent the map number, the habitat number, the largest cluster size, the largest cluster edge, the fractal dimension of the largest cluster, radius mean squared (RMS) of the largest cluster, total number of clusters, total edges, area weighted average cluster size (Sav), total number of sites/cells of the current habitat type (S_freq), and the correlation length. This file is not overwritten with each RULE execution; rather, *new data is appended to the end of the file.*

RULE can also produce, if the user so chooses, one of three types of map files as described below. The **map files** are ASCII text, space-delimited files of whole numbers, representing the maps created within RULE.

Generated Map. The first is a file containing the map (or maps) created by RULE. Within a given map, each cell is numbered according to the habitat type. If multiple iterations are executed (and, hence, multiple maps are created), each "map" in the file is separated by a blank line. (NOTE: If the user is inputting, analyzing, and summarizing multiple maps in a single execution of RULE, each map within the file must be separated by a blank line, just as is done when RULE creates a map file.)

Cluster Size Map. In these maps, each cell is given a number based on the patch to which it belongs (e.g., all cells comprising the largest patch are labeled "1," etc.); the clusters are numbered consecutively according to size, from the largest to the smallest. A separate map is created within the file for each habitat type; thus, for an execution of RULE with multiple iterations and multiple habitat types, the number of maps within a single cluster-size map file equals (number of iterations) × (number of habitat types).

Cluster ID Map. This file contains a map in which, for each habitat type and each iteration, clusters are numbered in the order in which they are identified by RULE. As with the cluster-size map file, the number of maps in a cluster ID file equals (number of iterations) × (number of habitat types).

EXERCISES

EXERCISE 1
Simple Random Map(s) Analyzed with Three Different Neighborhood Rules

The purpose of this exercise is to become familiar with the methods for generating and analyzing patch structure in random maps and the effect of changing neighborhood rules. Your first step is to generate by hand a simple random map with rows and columns equal to 10 and $p = 0.5$. This exercise should be done with paper and pencil according to the following steps. (Alternatively, see the instructions for using spreadsheet software, listed *after* the "by-hand" instructions).

Instructions for generating and analyzing a simple random map "by-hand"

1. Use graph paper to create a grid with ten rows and columns.
2. Repeatedly flip a coin to determine the habitat type of each cell. If heads, then the habitat type equals 0. If tails, then the habitat type equals 1.
3. Analyze the map by coloring in all sites with habitat type = 1. (If you used Excel to do the first two steps, then color grid sites with random numbers ≤ 0.5).
4. Count the total number of colored cells, the total amount of edge, the number of clusters as defined by the nearest-neighbor rule, and the size of the largest cluster.
5. To calculate the total amount of edge, count, *for every cell equal to 1*, the number of neighboring cells equal to 0. Note that this means that any given cell containing a 0 may be counted more than once. In addition, note that the border of the map is *not* counted as edge. See Figure 9.3 for an example.
6. Repeat the analysis with the same map for the next-nearest-neighbor rule and the third-nearest-neighbor rule (Figure 9.1). (A hint to simplify the process of counting edges: Create a paper mask with holes in the shape of the three neighborhoods shown in Figure 9.1. Then, move this mask over each cell that is equal to 1, with the cell located in the center of the mask, and count the number of cells within the mask that are equal to 0.)
7. Record your results in Table 9.3 from the CD.

Instructions for using a spreadsheet to generate a matrix of random numbers

A matrix of random numbers can be generated by using spreadsheet software. To do so using Excel:

1. Open a new worksheet
2. Type this equation in the first cell: "=rand()"; this produces a single random number between the interval 0.0–1.0.
3. Copy this cell to a 10 × 10 grid of cells.
4. Print the resulting matrix and go to step 3 of the "by hand" directions.

Question 1. The generation of maps by hands is a tedious exercise that results in an inadequate sample size. Does the number of habitat sites of type 1 equal exactly 50% of the map? How many sites with habitat of type 1 touched the edge of the map? How many of these sites that touched the edge of the map would have adjoined another similar habitat site if the map size was increased? (HINT: See Gardner et al., 1987, for a discussion of cluster truncation effects.) How big would clusters be if the map size were increased?

Question 2. Combine your results with those of other students and statistically summarize (e.g., mean and standard deviation of total amount of edge, number of clusters, and the size of the largest cluster). What are the most reliable statistics (i.e., which ones have the lowest coefficient of variation)?

Question 3. Do you expect results from actual landscapes to be more or less variable than random maps?

EXERCISE 2
Landscape Connectivity

Next, you will use RULE to generate and analyze a series of random maps, investigating a variety of indices useful for characterizing landscape connectivity.

1. Use RULE to generate ten iterations of a simple random map of 128 rows and columns and a single habitat type with $p = 0.2$, using the nearest-neighbor rule for analysis. Save the output generated in the file rulerun.log by renaming the file—copy rulerun.log to a new file, ran02.log.

   NOTE: When creating a map with a single habitat type, the number of map classes is set to 1, even though the map technically contains two classes—the habitat of interest (type 1) and nonhabitat (type 0). Thus, when setting p for the single habitat type, the value of the nonhabitat type must be set first. The value of nonhabitat type, referred to in RULE as the "0th probability," is equal to $1.0 - p$. See Table 9.1 for an example.

2. Repeat step 1 three more times, setting the value of p for the single habitat type of each map to 0.4, 0.6, and 0.8, respectively, renaming each rulerun.log file, for a total of four sets of analyses.

3. Repeat steps 1 and 2 two more times using the next-nearest neighbor rule and the third-nearest-neighbor rule (a total of 12 sets of runs of RULE). The file **quick.bat** shows you how to perform these runs by one simple command, but you must create the 12 script files with a text editor; an example of the format and elements of a script file can be found in **random.scr**. Alternatively, you may perform these simulations interactively, but remember that each time RULE is run, it overwrites the rulerun.log file! Also, note that turning the visualization off significantly increases the speed with which Rule performs the analysis (although you should run RULE at least once with visualization activated to get some idea of the way the program operates).

4. Compare the results for the different neighborhood rules using the indices generated by RULE as a function of p and explain the trends you find in the results. (HINT: graphs are useful for this and may be produced easily using spreadsheet software.)

Question 4. A graph of results (particularly the correlation length) should illustrate the effects of critical thresholds. What are critical thresholds, and why are they important in understanding the effects of landscape change?

Question 5. Are critical thresholds independent of neighborhood rules?

Question 6. Do real landscapes have thresholds similar to those of random maps? If so, will these be at the same value of p as random maps?

EXERCISE 3
Effects of Fragmentation on Pattern

Two factors affect the spatial patterns produced by the process of landscape fragmentation. The first factor is the loss of habitat (i.e., reduction in p), and the second is the degree of contagion or autocorrelation among sites. Multifractal maps allow the effects of autocorrelation to be examined by changing a single parameter, H. In this exercise, you will change p and H to vary spatial pattern in multifractal maps. Here you will perform a factorial experiment with three levels of H and p (12 sets of simulations) using multifractal maps following these steps:

1. Generate four sets of multifractal maps with p for a single habitat type set to 0.2, 0.4, 0.6, and 0.8, respectively (don't forget, the "0th probability" required by RULE = $1.0 - p$). In addition to these values for p, use the following parameters in RULE:
   (a) Set the number of levels to 7 (this produces a 128 × 128 map).
   (b) set $H = 0.1$.
   (c) Choose "no" for wrapping the map.

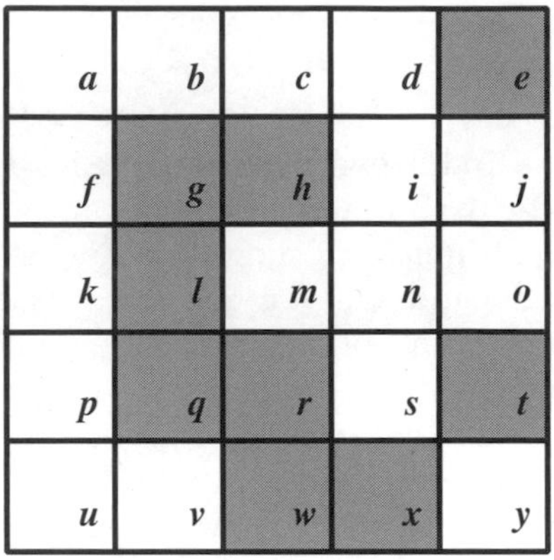

FIGURE 9.3
Sample grid to illustrate how to calculate total edge. Cells in gray represent grid
sites equal to 1, while nonshaded grid sites equal 0. Total edge is calculated by
counting, for each cell containing a 1, the number of adjacent sites containing 0.
Edges adjacent to the border of the map are not counted. Thus, for this map, and
using the nearest-neighbor rule, edge is calculated as follows: cell *e* = two edges (*d*
and *j*); cell *g* = two edges (*b* and *f*); cell *h* = three edges (*c*, *i*, and *m*); cell *l* = two
edges (*k* and *m*); cell *q* = two edges (*p* and *v*); cell *r* = two edges (*m* and *s*); cell *t* =
three edges (*o*, *s*, and *y*); cell *w* = one edge (*v*); cell *x* = two edges (*s* and *y*); total
edges = 19. Note that if other neighborhood rules are used, additional neighboring
sites containing 0's must be counted (e.g., diagonal cells are considered with the
next-nearest-neighbor rule; for this map, the number of edges would be equal to 36).

(d) Use neighborhood rule 2 (next-nearest-neighbor rule).
(e) Set the number of iterations to 10.
As in Exercise 2, you should perform at least one iteration with visualization turned on to see how RULE behaves differently with multifractal maps, particularly with different values of *H*. Also note that analyses for multifractal maps take considerably more time than for random maps, especially with higher values of *H*.

2. Repeat steps 1 setting *H* equal to 0.5 and 0.9.

3. Compare the results from the analysis in RULE as a function of *p* and *H*. In addition, compare and contrast the random maps from Exercise 2 (i.e., those using neighborhood rule 2) with the fractal maps created here in terms of pattern observed in each case.

4. In terms of the indices generated by RULE, consider how the value of *H* affects the observed landscape patterns. How do the results compare with those generated for random maps?

Question 7. Which results from the multifractal maps are most like those from analyses of simple random maps? Why? (HINT: Think about how the value of *H* changes the qualitative appearance of multifractal maps, and how this relates to trends in landscape indices.)

Question 8. What statistical output produced by RULE is the single best descriptor of differences in pattern between random and multifractal maps?

Question 9. How do changes in the value of *H* affect pattern? That is, which indices are most strongly affected by changes in *H*, and how would you explain these differences?

Question 10. Choose your favorite multifractal map pattern (i.e., combination of *p* and *H*, with the same random number seed). Generate a map with those values of *p* and *H* but with rows and columns equal to 64 (i.e., half the size used in this exercise); this requires setting the number of levels to 6 instead of 7. Describe and explain the patterns produced with respect to indices generated by RULE. How does the map differ from the 7-level (128 × 128) case?

OPTIONAL EXERCISES

EXERCISE 4
Neutral Models and Real Landscapes

Now you will explore the use of neutral models in forming hypotheses regarding structural patterns observed in actual landscapes. This exercise requires a bit more thought about the processes that influence structural patterns observed in actual landscapes versus those used to generate random computer maps. A sample 128 × 128 map, taken from the Patuxent River watershed in central Maryland, has been provided for use in this exercise. The Patuxent watershed lies within the piedmont and coastal plain of Maryland, just west of Chesapeake Bay. It is a highly urbanized area, encompassing the corridor between Baltimore and Washington, DC, as well as the state capital of Annapolis. Because of the high population density within the region, most forested land is highly fragmented.

1. Import the Patuxent map (**patgrid.map**) into RULE (i.e., choose "Input existing map file" as the first option in RULE). This particular section of the Patuxent watershed lies approximately 25 kilometers southeast of Washington, DC, and 10 kilometers west of Chesapeake Bay within a relatively less disturbed region just off the banks of the Patuxent River. The map size is 128 rows and columns, with forested areas represented by 1's and nonforested regions by 0's; thus, you will set the number of map classes to 1.

2. Choose a neighborhood rule to use and be prepared to justify your response. How does the analysis change with alternative neighborhood rules? (*NOTE:* When analyzing patgrid.map in RULE, be sure to remember to set the number of iterations to 1—otherwise, RULE will try to look for more than one map and will then give an error statement.)

3. Make sure you rename both the rulerun.log *and* the cfd_scr.dat files, since you will need them both for this exercise.

4. Formulate a hypothesis regarding which type of map would best serve as a neutral model for comparison with the Patuxent watershed map. Justify your answer, stating clearly what led you to this hypothesis.

5. Using what you have learned from previous exercises, run ten iterations of your neutral models in RULE and compare your results with those generated from the analysis of the Patuxent map.

Question 11. What proportion of the map is forested? Using the data from the cfd_scr.dat file, calculate the mean patch size. What is the difference between the largest cluster size, the area weighted average cluster size (Sav), and the mean patch size? Does the neighborhood rule used in the analysis change the results, and if so, how? How do you determine the appropriate neighborhood rule?

Question 12. Do the results for the Patuxent map fall within the confidence limits of those for your neutral model? Is your hypothesis supported or refuted by your results? What do your results tell you about processes that influence patterns in actual landscapes?

Advanced Question 13. How does the frequency distribution of cluster sizes of the Patuxent map compare with the frequency distribution of the neutral model? RULE produces a data set during each run called cfd_scr.dat, which contains information for each cluster identified including size, amount of edge, fractal dimension, and radius mean squared. This information can be imported to a spreadsheet, sorted by size, and the distributions compared. A plot of the cumulative frequency distribution (cfd) of cluster sizes can be produced by importing the second column of the cfd_scr.dat file into Excel, sorting this vector, and plotting the sorted numbers (on the *x*-axis) against the cumulative frequency (*y*-axis). For the cfd of each map, compare SAV (the area weighted average cluster size) with the mean and median of the cfd. Matching cfd's for real and artificial maps is one of the most difficult (and perhaps most critical) tests of the adequacy of a neutral model.

EXERCISE 5
Using ArcInfo to Create a Gridded Map

It is a relatively simply matter to use a GIS, such as ArcInfo, to clip portions of maps from GIS layers, import into RULE, and compare with random models. A long-term project might consist of generating random maps with more than one habitat type. The following directions provide detailed instructions for clipping gridded maps from ArcInfo data layers.

Instructions for clipping portions of a map using ArcInfo Grid, for export into RULE

1. Run the DESCRIBE command on the grid you wish to use (e.g., "DESCRIBE <gridname>"). Note the cell size and the units in which the grid is registered.

2. Start ArcEdit and type "GRIDEDIT EDIT <gridname>", then "DRAW." Note that the display environment (using the "DISPLAY 9999" command) must be set up for the grid to be displayed.

3. Find the portion of the map you are interested in clipping. Point the cursor to the lower left-hand corner of that portion; note the x and y coordinates listed at the bottom of the display window. Exit ArcEdit.

4. For the GRIDCLIP command, you will need to know the lower-left and upper-right coordinates of the portion in which you are interested. The former you have already obtained; the latter can be determined in the following way. First, determine the size of the map you wish to create (for example, a 128 × 128 map). Multiply the cell size of the grid (e.g., 30 meters) by the number of cells you want in the output map (e.g., 128) and add this number to the lower-left x and y coordinates. This will give you the upper-right x and y coordinates, respectively.

5. At the "Arc:" prompt, type "GRIDCLIP <gridname> <newgridname> <lower-left-x> <lower-left-y> <upper-right-x> <upper-right-y>". The new grid will include the clipped portion in which you are interested.

6. You must then turn the grid into an ASCII file: type "gridascii <newgridname> <ASCII-filename>".

7. Open the ASCII file in a word processor and delete the first several lines, ending with "NODATA_value -9999": this is the header information describing the map, and it is not necessary for use with RULE. Then, change all NODATA values (usually -9999) to zero. Also, you will probably wish to change the grid values to a more convenient set of integers—for example, changing all forest pixel values to 1, agriculture values to 2, and so on. To analyze maps in RULE by habitat type, you must have the habitats classified with consecutive integer values, starting with 1.

ACKNOWLEDGMENTS

Preparation of the manuscript was funding in part by funds from the U.S. Environmental Protection Agency under contract R819640 to the Center for Environmental Studies, University System of Maryland, and by a grant from the National Science Foundation (EAR-9506606). Support for S. Walters was provided by the U.S. EPA (in conjunction with the National Science Foundation), Office of Research and Development, National Center for Environmental Research and Quality Assurance (R82-4766-010). This paper is scientific contribution #3210 from the University of Maryland Center for Environmental Science, Appalachian Laboratory.

BIBLIOGRAPHY

Note. An asterisk preceding the entry indicates that it is a suggested reading.

GARDNER, R. H. 1999. RULE: A program for the generation of random maps and the analysis of spatial patterns. In J. M. Klopatek and R. H. Gardner, eds. *Landscape Ecological Analysis: Issues and Applications*, Springer-Verlag. New York, pp. 280–303.

*GARDNER, R. H., B. T. MILNE, M. G. TURNER, AND R. V. O'NEILL. 1987. Neutral models for the analysis of broad-scale landscape pattern. *Landscape Ecology* 1:19–28. This paper introduces the concept of using neutral landscape models to explore the effects of ecological processes on landscape pattern. Comparisons between randomly generated maps and USGS land-use data maps reveal more aggregated clusters in the latter, suggesting the importance of processes such as disturbance in determining landscape structure. The existence of critical thresholds in landscape connectivity was also explored in the analysis of neutral landscapes.

GARDNER, R. H., R. V. O'NEILL, AND M. G. TURNER. 1993. Ecological implications of landscape fragmentation. In S. T. A. Pickett and M. J. McDonnell, eds. *Humans as Components of Ecosystems: Subtle Human Effects and the Ecology of Populated Areas*. Springer-Verlag, New York, pp. 208–226.

*MCINTYRE, N.E., AND J.A. WIENS. 1999. Interactions between habitat abundance and configuration: Experimental validation of some predictions from percolation theory. *Oikos* 86:129–137. This paper describes the application of neutral model predictions to a field experiment exploring the movements of tenebrionid beetles. Experiments revealed the predominant impact of habitat abundance on movement behaviors, especially when habitat is sparse, as neutral model predictions suggest. The study illustrates the use of landscape approaches in conservation endeavors, and the importance of explicitly considering spatial pattern in habitat management efforts.

PEARSON, S. M., AND R. H. GARDNER. 1997. Neutral models: Useful tools for understanding landscape patterns. In J. A. Bisonnette, ed. *Wildlife and Landscape Ecology: Effects of Pattern and Scale*. Springer Verlag, New York, pp. 215–230.

PLOTNICK, R. E., AND R. H. GARDNER. 1993. Lattices and landscapes. In R. H. Gardner, ed. *Lectures on Mathematics in the Life Sciences: Predicting Spatial Effects in Ecological Systems*, Vol. 23. American Mathematical Society, Providence, Rhode Island, pp. 129–157.

STAUFFER, D., AND A. AHARONY. 1992 *Introduction to Percolation Theory*. Taylor & Francis, London.

SAUPE, D. 1988. Algorithms for random fractals. In H.-O. Peitgen and D. Saupe, eds. *The Science of Fractal Images*, Springer-Verlag, New York, pp. 71–113.

WITH, K. A., R. H. GARDNER, AND M. G. TURNER. 1997. Landscape connectivity and population distributions in heterogeneous environments. *Oikos* 78:151–169.

*WITH, K. A., AND A. W. KING. 1997. The use and misuse of neutral landscape models in ecology. *Oikos* 79:219–229. The applications of neutral models to problems in landscape ecology are reviewed here. The authors discuss the appropriateness of using neutral landscape models to infer relationships between pattern and process. The paper also discusses misuses of neutral models, describing analyses that mistakenly conclude a direct causal connection between pattern and process.

SCALE DETECTION USING SEMIVARIOGRAMS AND AUTOCORRELOGRAMS

MICHAEL W. PALMER

OBJECTIVES

Landscape ecology differs from most other branches of ecology in that it explicitly involves space. Therefore, one of the goals of landscape ecology is to describe spatial variation. The purpose of this exercise is to

1. introduce two tools for describing this variation: semivariance and auto-correlation; and
2. give students experience creating and interpreting semivariograms and autocorrelograms

In this lab, you will collect field data from quadrats arranged along a transect (or alternatively, you will use supplied data). You will then calculate and graph semivariograms and autocorrelograms using a spreadsheet, and you will use these graphs to determine how spatial patterns vary as a function of scale in your system. For this lab, you will need access to a spreadsheet program (such as Excel) and the file **vario.xls** included on the CD. If you choose the fieldwork option, you will also need two 100-meter measuring tapes and one 1 × 1-meter sampling quadrat.

INTRODUCTION

Nature is intrinsically variable, and the evolution and ecology of all organisms are contingent on such variation. Landscape ecology is concerned not only with the magnitude of this variation, but also with its geometry. Most patterns in nature are far more complex than the simple polygons and curves of Euclidean geometry. For example, forest edges are rarely straight lines, animal home ranges are not rectangles, and trees are not cones. Therefore, we need special methods to describe the shape of nature.

The discipline of spatial statistics has diversified and matured (see Cressie, 1991; Bailey and Gatrell, 1995), and it is not possible here to give a full summary of the wealth of methods available. Instead, the purpose of this exercise is to describe two different methods for characterizing variation in a variable as a function of position in the landscape. This variable could be a soil nutrient, a measure of vegetation height, an index of species composition, or anything else of interest. In spatial statistics, we term variables with known locations **regionalized variables,** and we label them z (so as not to confuse them with x and y, typically reserved for the spatial coordinates, or for independent and dependent variables, respectively). The two methods covered in this exercise are variography, which is part of the discipline of geostatistics (see Isaaks and Srivastava, 1989), and autocorrelation, which is derived from the familiar correlation coefficient (Sokal and Rohlf, 1981). Recall that the correlation coefficient, r, is a number that varies between -1 and $+1$ and reveals the nature of the relationship between two variables. It is close to -1 for two variables that are strongly negatively related, close to 0 for unrelated variables, and close to $+1$ for positively related variables. In contrast, variography is derived from the *variance*, which must be a positive number but can otherwise take any value.

One of the most important properties of almost all regionalized variables is **spatial dependence.** Spatial dependence (as assessed by **spatial autocorrelation,** or the tendency of a random variable to be correlated with itself at finite distances) means that a variable measured at one location *depends*, in one way or another, on the same variable measured at a different location. Spatial dependence arises for a number of different reasons, but let us consider two examples.

If you examine mean annual temperature as a function of position on the globe, you will note that (with many important and interesting exceptions) there is a gradient from warm temperatures at the equator to cool temperatures at the poles. If you have two sites that are almost at the same latitude, they will have similar temperatures. On the other hand, two sites that are on different latitudes will have different temperatures, and the amount of the difference in temperature will be positively (and gradually) related to the difference in latitude. **Spatial dependence** occurs when information available at one location allows you to infer information about the other location. Another example is in a savanna landscape where widely spaced trees provide islands of shade in an otherwise sunny landscape. Two sites that are centimeters apart are likely to have a similar amount of sunshine. However, two sites that are several meters apart may, or may not, have similar amounts of sunshine—a lot depends on the size and spacing of trees. If the sites are hundreds of meters apart, you may not be able to predict the sunlight regime very well. So in this case, we have spatial dependence at fine scales, but not necessarily at coarse scales. Also, unlike the example of global temperature, our regionalized variable consists of fairly discrete *patches* of sun and shade.

The first column of graphs in Figure 10.1 displays a variety of made-up regionalized variables with identical means and variances. These hypothetical variables have been constructed to illustrate the diversity of patterns that could potentially be found in nature. Note that regionalized variables can consist of a variety of features such as patches (i.e., homogeneous regions), noise (random, independent variation), random walks (a random walk is when a value at a given location equals the value at an adjacent location, plus or minus a small random number), or some combination of these. Also, note that the different variables behave differently as a function of scale. For example, patches can be large, small, or intermediate. Stretches of linear behavior can also be large, small, or intermediate. Also, noise can operate at any scale. If the graphs in Figure 10.1 were based on real data, we would seek biological explanations for the different scales. Such explanations might involve the size and shape of underlying geomorphology, the average size of plant clones, the average size of a natural disturbance, the home range size of the dominant mammal species, or the average farm size.

Except for variable A in Figure 10.1, there is some spatial dependence. That is, nearby locations are, *on average*, more similar than distant locations. Since similarity typically decreases as a function of distance of separation, we also call this phenomenon **distance decay**. Distance decay has important consequences for living things. For example, if soil conditions are very similar at nearby locations (as for variables C and F in Figure 10.1), then natural selection *might* favor plants with short dispersal distances. If, on the other hand, soil conditions were spatially unpredictable (as for variable A), a long dispersal distance *might* be advantageous. Similarly, the foraging behavior of animals, the growth of plant roots, the spread of fire, the flow of water, and the behavior of many other ecological phenomena all depend on the nature of distance decay in environmental factors.

In statistics, spatial dependence has both desirable and undesirable attributes (Legendre, 1993). It means that one can predict variables (to some degree) based on geographic location, which can aid in mapping the environment. However, spatial dependence also violates the standard statistical assumption of independent observations (even if samples are randomly located). Thus, unless specifically corrected for, many statistical methods are invalid if your data exhibit distance decay. Fortunately, there are tools to evaluate the *degree* and the *scales* of spatial dependence. The two tools we introduce in this laboratory exercise are the semivariogram and the autocorrelogram.

Variography

Variography is the discipline of using semivariograms (and related graphs such as covariograms) to uncover the degree to which the variance in a regionalized variable depends on distance (Rossi et al., 1992). The geographic distance between two samples is termed the **spatial lag**.

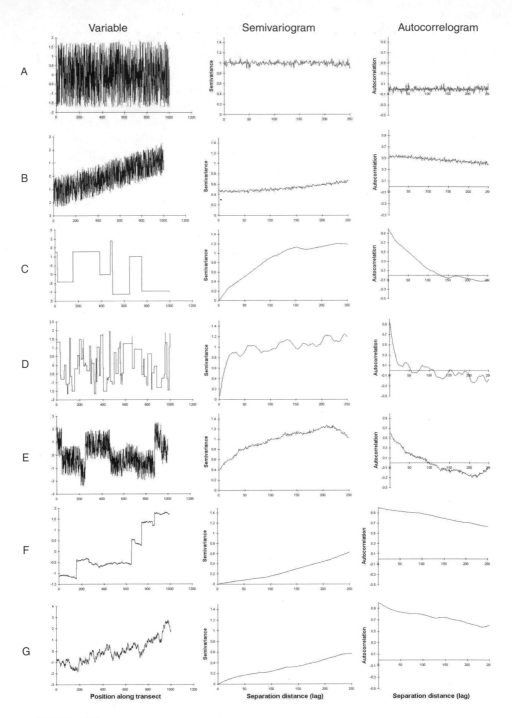

	Variable	Semivariogram	Autocorrelogram

FIGURE 10.1

Seven artificial regionalized variables (column 1) as a function of position along a transect, along with their corresponding semivariograms (column 2) and autocorrelograms (column 3). All variables have identical means and variances. The variables can be described as follows: (A) pure noise; (B) fine-scale noise superimposed on a linear trend; (C) large patches; (D) small patches; (E) noise superimposed on large patches; (F) patches with "drift" in their mean values, plus fine-scale noise; (G) random walk.

Recall that the variance is the square of the standard deviation and is a measure of the spread or variation of data. The word **semivariance** is derived from "half of the variance," and indeed it is a measure of the variance of the regionalized variable, z. But what is special about the semivariance is that it changes as a function of distance.

The semivariance is computed as follows:

$$\gamma(h) = \{\Sigma[z(i) - z(i + h)]^2\} / 2N(h)$$

where $\gamma(h)$ is the semivariance of a **lag** of distance h, $z(i)$ is the value of a regionalized variable z at location i, $z(i+h)$ is the value of z at a location separated from i by lag h, and $N(h)$ is the number of pairs of points separated by lag h. The summation is over all pairs of points separated by distance h. In plain English, the semivariance is half of the average squared difference of all pairs of points separated by a given distance. A semivariogram is a plot of semivariance versus the lag distance. As with the variance, the semivariance cannot be less than zero, but it is not bounded on the top.

An idealized, hypothetical semivariogram is given in Figure 10.2. Since the semivariance is directly related to variance, a high value indicates high variation, and a low value indicates low variation. Almost always, variance increases as a function of lag distance. In other words, the larger the area you study, the more variable your conditions are. It is important to reiterate that the lag distance is not the same as the distance from the origin or starting point. Rather, it is calculated for all pairs of points (Figure 10.3).

At distances less than R (the **range**), we have spatial dependence (Figure 10.2). That is, closer samples are more similar than distant samples. At distances of at least h, we have spatial independence; therefore, samples separated by longer distances would be valid for conventional statistics. Any area with linear dimensions of at least R would have as much variance as the land-

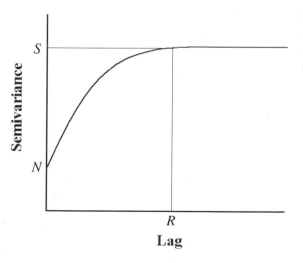

FIGURE 10.2
An idealized semivariogram.
N = nugget, R = range,
S = sill.

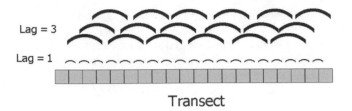

FIGURE 10.3
A 20-quadrat-long transect illustrating all of the pairs of points separated by two selected lags: 1 and 3.

scape as a whole. A horizontal asymptote at distances greater than R is known
as the **sill.** The sill indicates the amount of "background" variation.

For a very smooth, regionalized variable, two samples that are infinitesi-
mally close to each other may have almost identical values. Elevation of the
ground surface almost always behaves this way. However, most variables are
not so smooth, if for no other reason than measurement error. Such unre-
solved variation at very fine scales is termed the **nugget effect,** and is indicated
by N (Figure 10.2). The term derives from the original use of variography in
gold mining: at fine spatial scales, you either find the gold nugget or miss it.
Soil variables such as pH or nutrient concentrations typically have very high
nugget effects.

The semivariogram specifically address how variance increases as a function of scale. Although it can only describe *patterns*, we often hope to infer
the *processes* that generate such patterns. If we find a distinct range, or even
a pronounced inflection in the semivariogram, we suspect that there are different processes operating at different scales. For example, if we discover that
the range equals approximately 10 meters, we need to seek an underlying
process that operates on a scale of 10 meters. In a forest, this scale could represent the average canopy gap size or the average size of a canopy tree crown.
In the arctic tundra, the range could represent the average size of a permafrost
polygon. Of course, you are never guaranteed to actually find a range. It is
possible (and indeed, likely) that variation in nature increases continuously as
a function of scale.

The second column of Figure 10.1 shows the semivariograms for the hypothetical regionalized variables. Since very few pairs of points represent very
far distances, it is usually not advisable to plot $\gamma(h)$ for large h. A general rule
of thumb (adopted here) is to plot only up to half of the maximum distance
between samples. The bumps and wiggles at such far distances in the semivariograms are due to chance variation in the data, and not to the underlying process generating the patterns. Note that only three of the variables (C,
D, and E) have semivariograms remotely resembling those in Figure 10.2. The
range of variable C (≈ 150 units) is much larger than the range of variable D
(≈ 25 units), which reflects the differences in patch size. Variable E seems to

have an inflection at 50 meters, which marks the difference between noise within patches, and the differences between patches.

The semivariograms of variables B, F, and G are continuously increasing functions. Therefore, there is spatial dependence at broad spatial scales. Three variables (A, B, and E) have much fine-scale noise, and hence a substantial nugget effect. Note that variable F has very little fine-scale noise, and hence has a negligible nugget. Variable A represents pure noise and hence pure spatial independence. For such variables, the nugget equals the sill.

Autocorrelation

Autocorrelograms are plots of the correlation coefficient, r, as a function of lag:

$$r(h) = \text{corr}[z(i), z(i + h)]$$

It is called "auto"-correlation because the variable is correlated with itself. Autocorrelation can take values from -1 to $+1$, although for most applications positive values are most common. In situations with distance decay, autocorrelograms are declining functions and often look like upside-down semivariograms (third column of Figure 10.1). If there is little fine-scale noise, the y-intercept will be close to 1.0. In situations in which the semivariogram displays a nugget effect, the y-intercept of the autocorrelogram will be less than 1. An autocorrelation of 0 means there is no spatial predictability; this is related to the concept of the sill.

This describes only the simplest kind of autocorrelogram. More complex (and usually more appropriate) ways to calculate autocorrelograms, as well as testing their statistical significance, are described by Legendre and Fortin (1989), Bailey and Gatrell (1995), and Legendre and Legendre (1998). Autocorrelograms are also used in the analysis of change through time (also known as "time series").

Comparing Autocorrelation to Semivariance

The interpretation of autocorrelograms is very similar to that of semivariograms, so the choice between them is largely a matter of taste. Since the correlation coefficient is a dimensionless number (i.e., it is standardized), autocorrelograms are useful in comparing variables with different units (e.g., plant density and soil calcium). Semivariance has a dimension of units squared (so if the regionalized variable is in parts per million or ppm, semivariance is in ppm^2). Thus, it is useful in comparing different commensurate variables or (more commonly) the same variable in different locations. However, semivariance can be standardized for comparing variables measured in different units (see Rossi et al., 1992).

Since it is derived from the correlation coefficient, autocorrelation is closely related to classical statistical theory. Variography, on the other hand, is a branch of geostatistics. This discipline was largely developed for the mining industry to help predict the locations of mineral deposits. Variography is a precursor to geostatistical interpolation (for mapping) or "kriging" (see Isaaks and Srivastava, 1989) and to fractal geometry (Burrough, 1983; Palmer, 1988).

EXERCISE

DATA COLLECTION

Option 1: Field Exercise Using Vegetation Height

1. Choose a field site in which the maximum height of the vegetation is about 2.5 meters or less, and in which it is possible to fit a 200-meter transect. If the site is large enough, randomly choose a starting location and compass direction. If it is too small for this, choose an appropriate direction but randomize the starting point, so you are not biased by particular plants.

2. Extend two (or more) 100-meter tapes end to end along the chosen compass direction. Ideally, you would do this with a surveyor's compass or level. It is crucial that the transect be as straight as possible and not influenced by the vegetation!

3. Beginning at 0 meters, establish a 1 × 1-meter quadrat (this can consist of three meter sticks plus the meter tape as the fourth boundary).

4. Within this plot, measure and record the height of the tallest plant.

5. Now repeat the process with an adjacent plot at 1 meter, then at 2 meters, and continue to the end of the transect.

If you do not have the luxury of a large enough field site, it is possible to perform this exercise with smaller contiguous quadrats. Also, the regionalized variable need not be height: you can perform this same exercise using stem density, biomass, species richness, ordination scores, percent cover of bare soil, elevation, percent sunlight, soil parameters, and more. It may be possible to derive a regionalized variable from a map or a remotely sensed image, but be aware that data on such images *may* have already been "smoothed" or interpreted for ease of display, and hence your analyses would be inappropriate. Regardless of the overall length and quadrat size, try to have at least 200 quadrats in your transect: spatial analyses typically require large sample sizes. Another option is to split the class up into two or more groups, with each group studying either a different regionalized variable along the same transect, or the same variable in a different vegetation type.

Option 2: Using Provided Data Sets

Some example data sets are provided on the accompanying CD-ROM, in case there are no opportunities for collecting new data. The file is entitled **vario.xls** and contains a worksheet with three different **"example data sets."**

DATA ANALYSIS

You will analyze the data using a spreadsheet. The example given here is for Microsoft Excel, but similar commands exist in other spreadsheets. Before beginning this exercise, review absolute and relative cell references, how to graph data, as well as the following Excel functions: OFFSET, SUMXMY2, CORREL. Make sure that automatic calculation of formulas is in effect (this is the most usual default; it means that the results will be continuously updated. Check under Tools–Options–Calculation–Automatic.).

1. Enter your data in the blank worksheet labeled **Vegetation Heights.** The following description assumes that the transect is 200 quadrats long; if not, substitute "200" with the correct number.

2. In row 1, label columns **A** through **F** as follows:

A	B	C	D	E	F
POSITION	VALUE	LAG	SEMIVARIANCE	LAG	AUTOCORRELATION

3. In column **A,** fill rows **2** through **201** with the numbers 1 through 200. One quick way to do this is to put "1" in cell **A2,** and then put the formula " = **A2 + 1**" in cell **A3.** Then copy the contents of **A3** and paste them into cells **A4** through **A201.** Since there were no dollar signs ($) in the original formula, the cell reference of "**A2**" is copied as a relative location. Therefore, each one of the cells will equal the cell above it plus 1.

4. In column **B,** fill rows **2** through **201** with the data you collected (or copied from the provided data sets) in the correct spatial sequence.

5. In column **C,** fill rows **2** through **101** with the numbers 1 through 100. (Recall the rule of thumb that it is best not to plot semivariograms for more than half of the maximum lag distance). Repeat this for column **E.**

Two different ways to calculate the semivariance follow. Method 1 is conceptually easier, but method 2 is less labor intensive. Therefore, read and understand method 1, but use method 2. It is important to keep in mind that in a transect Q units long, the number of pairs of quadrats separated by a given lag distance h will equal $Q - h$. In our example, there are 199 pairs of quadrats separated by 1 meter, 198 separated by 2 meters, and 1 pair separated by 199 meters (i.e., the first quadrat and the last quadrat). Before continuing, review the earlier equation for semivariance on page 133.

Semivariance method 1

1. In cell **D2**, put the formula "=SUMXMY2(B2:B200,B3:B201)/ (2*(200 − 1))." The formula SUMXMY2 means "sum of (x minus y) squared." The two selected blocks (**B2:B200**) and (**B3:B201**) are actually the same data, but shifted by a lag of 1 unit. The denominator is two times the number of pairs of points separated by distance h. It is, of course, possible to put the number 198 in the denominator, but writing the formula out often helps with troubleshooting.

2. In cell **D3**, write the formula "=SUMXMY2(B2:B199,B4:B201)/ [2*(200 − 2)]." This is the semivariance for a lag of 2. The formula for **D4** should be:"=SUMXMY2(B2:B198,B5:B201)/[2*(200 − 3)]." Continue filling in column **D** until you reach a lag of 100.

Semivariance method 2

1. Instead of typing in a unique formula for each cell of column **D**, it is more time-efficient to type in a generic formula in cell **D2**:

$$\text{"=SUMXMY2(B\$2:OFFSET(B\$2,200 − C2 − 1,0),OFFSET(B\$2,C2,0):B\$201)/(2*(200 − C2))".}$$

This is precisely the same formula as in method 1, except for how we specify addresses. The dollar signs ($) before the row means a reference to that exact row, no matter where you copy and paste the formula. OFFSET returns a new cell address and has three arguments: a cell address, the number of rows of separation, and the number of columns of separation. Therefore, the block **B$2:OFFSET(B$2,200 − C2 − 1,0)** refers to a column of data beginning at cell **B2** and ending (199 − **C2**) cells below **B2**. Since cell **C2** indicates a lag of 1, the column of data will be the same as **B2:B200**, as desired. The second block, OFFSET(**B$2,C2,0):B$201**, means a block beginning at **C2** below **B2**, and ending at **B201**. This will be the same as **B3:B201**. The denominator of the equation will equal 2*(200 − 1).

2. Copy cell **D2** and paste it into cells **D3** through **D101**. Note that when you do so, the formula remains identical in all cells *except* that the reference to the lag, column **C**, changes. Thus, the formula will return the semivariance for whatever lag is indicated in the same row of column **C**.

You can calculate autocorrelation by similar methods to those described earlier for semivariance.

Autocorrelation method 1 (Read and understand this method.)

1. In cell **F2**, type: "=CORREL(B2:B200,B3:B201)". This will return the correlation between the variable and itself, with a lag of 1.

2. In cell **F3**, type "=CORRELL(B2:B199,B4:B201)" for a lag of 2, and continue filling column **F** until lag 100.

Autocorrelation method 2 (Use this method.)

1. Following the same reasoning as method 2 for the semivariance, type the following in cell **F2**: "=CORREL(B$2:OFFSET(B$2,200 − E2 − 1,0),OFFSET(B$2,E2,0):B$201)."
2. Copy this formula and paste it into cells **F3** through **F101**.

Before proceeding, make sure to save your results.

RESULTS

Using your spreadsheet program, create the following plots:

1. Vegetation height as a function of transect position
2. Semivariance as a function of transect position
3. Autocorrelation as a function of transect position
4. Create a semivariogram as follows:
   (a) Plot the semivariance as a function of lag. The data will be in the block **C2:D101**.
   (b) Label the x-axis "lag (meters)," and the y-axis "Semivariance."
   (c) Drag the graph immediately under the graph of the raw data.
5. Create an autocorrelogram as follows:
   (a) Graph the data in **E2:F101** and drag the graph under the semivariogram.

(b) Label the axes appropriately.

HINT: Make an XY (scatter) plot, double-click on any point in the graph, then select Format Data Series–Patterns–Line–Automatic.

Interpretations of the Graphs and Rules of Thumb. As with any bivariate (two-variable) graph, the scaling of the y-axis relative to the x-axis should not affect our interpretation, but it often does. A short, long graph often appears less "noisy" than a tall, narrow one. It is generally best to choose a scaling relatively close to 1:1 (that is, square), or at most 1.5:1 or 1:1.5. Of course, there may be exceptions (e.g., if one wants to display the results of numerous transects, one graph on top of the other, it might be useful to have them short and long).

For semivariograms, it is conventional and advisable for both the x-minimum and the y-minimum to be zero. The x-minimum should be zero for the autocorrelogram, but a case can be made that the y-minimum and the y-maximum should be −1.0 and 1.0, respectively. If part of the goal of the research is to compare the results from different transects, x-axes and y-axes of the same kind of plot should be scaled identically.

The plot of vegetation height as a function of transect position will typically have some sort of broad-scale pattern, immediately detectable upon inspection, in addition to fine-scale variation. The details will vary markedly depending on the nature of your plant community. The fine-scale variation may be partially measurement error, but in most cases it is predominantly caused by natural variation. Increasing the number of samples (i.e., transect length) will not reduce the magnitude of this fine-scale variation.

The semivariograms and autocorrelograms will also have an overall shape, summarizing the spatial patterns of the community, as well as fine-scale variation. However, in contrast to the graph of height, increasing the sample size (transect length) will tend to decrease the finer-scale patterns. This means that we are increasingly confident that we have described how spatial pattern (variance or correlation) is related to scale (spatial lag).

Question 1. How does height behave as a function of distance along the transect? Is this generally consistent with your impression of the field site?

Question 2. Examine the semivariogram. Is there an identifiable nugget? Range? Sill?

Question 3. Does the regionalized variable (height) exhibit spatial dependence?

Question 4. Examine the autocorrelogram. Is there spatial autocorrelation?

Question 5. How would you describe the nature of your spatial variation? Does your pattern consist of patches? Noise? A dominant trend? Nested patterns of variation? Random walk? A random walk (also known as "drift") is when there is spatial dependence, but the difference between each number and the previous number is random. The term *random walk* derives from a plot of distance from the starting point as a function of time for an animal whose direction of movement is purely random.

Question 6. Is there periodicity in your data? How would you know this from the semivariogram or autocorrelogram? Note that both the semivariogram and the autocorrelogram can describe variance as a function of scale, but neither can completely summarize the *nature* of spatial variation (e.g., patches, gradients, or a combination). This is akin to the observation that variance does not fully describe the statistical distribution of data (e.g., whether it is normally distributed), and that the correlation coefficient does not fully describe the nature of the relationship between two variables (e.g., whether they might have a nonlinear relationship).

Question 7. Suppose a rodent species requires tall vegetation for cover. Does the nature of the spatial pattern you observe have implications for this species?

Question 8. Suppose a predator only hunts in relatively short vegetation. Does the nature of spatial variation have implications for foraging behavior?

Question 9. Can you think of any other biological ramifications of your results?

Question 10. If you collected data from more than one site or variable, how do their spatial patterns compare?

Question 11. How does your variable behave in comparison to the supplied data sets? The supplied data sets (on the page "example data sets" in the spreadsheet accompanying this laboratory) can be pasted into the data column (column **B**) in the worksheet "ready-to-go blank", and the semivariograms and autocorrelograms will be recalculated automatically (however, note that you may need to change the *y*-axis scaling on the graphs).

Question 12. Refer to Figure 10.1. Choose two or three of the variables and describe what natural phenomena might lead to those patterns.

Question 13. If you find spatial dependence, what does this imply for the use of conventional statistics?

Question 14. Are some spatial scales better than others for studying your system? Why or why not?

Question 15. In theory, regionalized variables are measured at points. However, you have measured them in a quadrat. What do you expect would happen to the semivariogram if you reduced the size of the quadrat?

Question 16. What is noise? Is it a useful concept?

OPTIONAL EXERCISES

EXERCISE 1
Correlation and Variation

As their names imply, the auto**correl**ograms and semi**vari**ograms stress correlation and variance, respectively. Therefore, they are likely to behave differently in data sets with different variance. In the supplied spreadsheet accompanying this exercise, locate a worksheet entitled "2 hypothetical variables." Examine the two variables carefully.

1. How do these variables differ?

2. How do you expect their semivariograms and autocorrelograms to differ?

3. Now copy one of the variables and paste it in the data column (column **B**) in the sheet labeled "ready-to-go blank." Examine the semivariogram and autocorrelogram. Now repeat with the second variable. Were you right in your answer to the previous question?

EXERCISE 2
Variography and Fractals

Plot your semivariogram on a double logarithmic scale. Do this by left-clicking on the *x*-axis of your semivariogram. Then right-click and choose **Format axis**. Select **Scale** and click **Logarithmic**. Repeat the same procedure for the *y*-axis.

1. Is the semivariogram a straight line? If so, we can say that the variable is *statistically self-similar*. This means that fine-scale patterns are

indistinguishable from scaled-down versions of broader-scale patterns. The concept of "self-similarity" is intrinsic to the study of *fractal geometry*. The fractal dimension D can be determined from the slope m of the log-log semivariogram with the formula $D = (4 - m)/2$. The interpretation of the fractal dimension is beyond the purpose of this chapter; see Burrough (1983) and Palmer (1988) for more details.

2. Are there multiple plateaus? If so, we have a hierarchy of spatial patterns. This would imply that we have distinctly different processes operating at distinctly different *scale domains*.

3. Would you predict that most spatial patterns in nature are self-similar, hierarchical, or neither?

FURTHER STUDY

This exercise only considered one-dimensional patterns. However, ecologists typically study spatial patterns in two dimensions. The same formulas for semivariance and autocorrelation hold, but the calculations are a bit more complicated. This is because distances no longer fall in discrete lag intervals. Therefore, we typically average semivariance over a certain range of lags. A further complication arises if the patterns are not *isotropic* (statistically the same in all compass directions). In such cases, we usually calculate different semivariograms and autocorrelograms for different directions.

For spatial analysis, sampling need not be in a perfectly sampled transect. It is perfectly legitimate for samples to have locations that are random, on interrupted transects and grids, or any other objective method. If the goal is to characterize spatial pattern, sample locations must be objective.

When samples are located at irregular intervals and/or in two dimensions, many of the spatial lags are not a simple multiple of the minimum spacing. We deal with this by creating "lag classes" (e.g., 0–1 meter, 1–2 meters, 2–3 meters, etc.) much in the same way as we would generate a histogram. Although there are no firm rules about how many pairs of points should fall within a lag class for an accurate semivariogram or autocorrelogram, a general rule of thumb is that it should be at least 80.

In this lab, we interpreted semivariograms and attempted to find the range, sill, and nugget by eye-balling. However, it is possible to use a curve-fitting procedure such as nonlinear regression to actually obtain estimates of these parameters (see Legendre and Legendre, 1998). However, it is not straightforward to make statistical inference on the shapes of these curves (e.g., to create confidence intervals for the range, sill, and nugget) because subsequent points on the semivariogram are not independent of one another. Nevertheless, such curve-fitting is an essential step for kriging (discussed next).

While we have only discussed univariate patterns, we are often interested in discussing bivariate or multivariate patterns. If so, we can use covariograms or cross-correlograms to determine whether the relationships between variables change as a function of spatial scale.

As previously mentioned, variography is often a precursor to a geostatistical interpolation procedure known as kriging (Hohn, 1988; Isaaks and Srivastava, 1989; Cressie, 1991). By interpolation, we mean that we estimate the value of a regionalized variable at an unsampled location, based on knowledge from sampled locations. Therefore, the most common product of kriging is a map (usually a contour map) of our variable of interest. Kriging performs best when the nugget is small relative to the sill and when the average distance between nearby samples is less than the range. See Legendre and Fortin (1989), Halvorson et al. (1994), Marinussen and van der Zee (1996), and Carroll and Pearson (1998) for examples of kriging in ecology.

For a simple analysis of spatial pattern along a transect, as in this lab, it is possible to perform basic calculations on a spreadsheet. However, a spreadsheet becomes cumbersome for more complex designs such as interrupted sampling or two-dimensionsional sampling, and for more complex analyses such as detection of anisotropy, significance testing, nonlinear curve-fitting, multivariate patterns, or kriging. Fortunately, a wide range of software exists, including C2D (http://www.exetersoftware.com/cat/c2d.html), GEO-EAS (ftp://math.arizona.edu/incoming/uix.geoeas/), GS+ (http://www.rockware.com/catalog/gs.htm), R Package (http://alize.ere.umontreal.ca/~casgrain/en/labo/R/), and Variowin (http://www-sst.unil.ch/geostatistics.html).

ACKNOWLEDGMENTS

I thank Steven Thompson, Sam Fuhlendorf, Anne Cross, Sophonia Roe, Marie-Josée Fortin, and four anonymous reviewers for comments on earlier versions. I especially thank Sam Fuhlendorf for allowing the use of his data, and José Ramón Arévalo for collecting new data for the lab. I thank Professor Eddy van der Maarel for the kind support of my stay at Växtbiologiska Institutionen, Uppsala University, Sweden.

BIBLIOGRAPHY

Note. An asterisk preceding the entry indicates that it is a suggested reading.

*BAILEY, T. C., AND A. C. GATRELL. 1995. *Interactive Spatial Data Analysis.* Longman Scientific & Technical, Essex, England. An excellent text on spatial statistics; includes some demonstration software.

BURROUGH, P. A. 1983. Multiscale sources of spatial variation in soil. I. Application of fractal concepts to nested levels of soil variations. *Journal of Soil Science* 34: 577–597.

*CARROLL, S. S., AND D. L. PEARSON. 1998. Spatial modelling of butterfly species richness using tiger beetles (Cicindelidae) as a bioindicator taxon. *Ecological Applications* 8:531–543. A good example of the application of spatial statistics in biodiversity studies

*COHEN, W. B., T. A. SPIES, AND G. A. BRADSHAW. 1990. Semivariograms of digital imagery for analysis of conifer canopy structure. *Remote Sensing Environment* 34:167–178. A good example of the use of variography in digital imagery.

*CRESSIE, N. A. C. 1991. *Statistics for Spatial Data.* Wiley Interscience, New York. A comprehensive and authoritative work on spatial statistics.

*HALVORSON, J. J., H. BOLTON JR., J. L. SMITH, AND R. E. ROSSI. 1994. Geostatistical analysis of resource islands under *Artemisia tridentata* in the shrub-steppe. *Great Basin Naturalist* 54:313–328. An example of ecological applications of geostatistics on a fine scale.

*HOHN, M. E. 1988. *Geostatistics and Petroleum Geology.* Van Nostrand Reinhold, New York. A good, clear text for the use of geostatistics for interpolation and mapping.

*HOULE, G. 1998. Seed dispersal and seedling recruitment of *Betula alleghaniensis*: Spatial inconsistency in time. *Ecology* 79:807–818. Another good example of geostatistics in fine-scale plant ecology.

*ISAAKS, E. H., AND R. M. SRIVASTAVA. 1989. *An Introduction to Applied Geostatistics.* Oxford University Press, New York. A popular text on the use of geostatistics for interpolation.

LEGENDRE, P. 1993. Spatial autocorrelation: Trouble or new paradigm? *Ecology* 74:1659–1673.

*LEGENDRE, P., AND M.-J. FORTIN. 1989. Spatial pattern and ecological analysis. *Vegetatio* 80:107–138. A good, brief introduction to spatial methods available to ecologists.

LEGENDRE, P., AND L. LEGENDRE. 1998. *Numerical Ecology*, 2nd English Edition. Elsevier, Amsterdam.

*MARINUSSEN, M. P. J. C., AND S. E. A .T. M. VAN DER ZEE. 1996. Conceptual approach to estimating the effect of home-range size on the exposure of organisms to spatially variable soil contamination. *Ecological Modelling* 87:83–89. A good applied example of the use of spatial statistics.

PALMER, M. W. 1988. Fractal geometry: A tool for describing spatial patterns of plant communities. *Vegetatio* 75:91–102.

*ROSSI, R. E., D. J. MULLA, A. G. JOURNEL, AND E. H. FRANZ. 1992. Geostatistical tools for modelling and interpreting ecological spatial dependence. *Ecological Monographs* 62:277–314. A good general introduction to variography, geared toward the ecologist.

*SCHLESINGER, W. H., J. A. RAIKES, A. E. HARTLEY, AND A. F. CROSS. 1996. On the spatial pattern of soil nutrients in desert ecosystems. *Ecology* 77:364–374. A good geostatistical study of soil nutrients.

SOKAL, R. R., AND F. J. ROHLF. 1981. *Biometry.* Freeman, New York.

4

DISTURBANCE DYNAMICS

Landscape disturbance, whether natural or anthropogenic, is an integral theme of landscape ecology because disturbances both create and are affected by spatial heterogeneity. Chapter 11, Landscape Disturbance: Location, Pattern, and Dynamics, examines several types of disturbance within the context of important concepts such as identifying the historic range of variability for a landscape and how scale influences perception of a disturbance-prone landscape. Chapter 12, Alternative Stable States, explores the theoretical construct of alternate stable states in the context of the spread of disturbance across landscapes using a simple spatial model of fire. Together these labs introduce a set of questions and approaches for addressing spatial aspects of disturbance.

CHAPTER 11

LANDSCAPE DISTURBANCE

LOCATION, PATTERN, AND DYNAMICS

MONICA G. TURNER, DANIEL B. TINKER, SARAH E. GERGEL, AND F. STUART CHAPIN III

OBJECTIVES

The causes, patterns, dynamics, and consequences of disturbances are major research topics in ecology and particularly landscape ecology (Romme and Knight, 1982; Risser et al., 1984; Pickett and White, 1985; Turner, 1987; Turner et al., 1989; Baker, 1989; Turner et al., 1997; Turner and Dale, 1998). Disturbances are also of tremendous importance in land and resource management, both for managing human activities (e.g., Hunter, 1993; Parsons et al., 1999) and for conserving resources and biodiversity in landscapes influenced by disturbance (e.g., Dale et al., 1998). In this lab, students will

1. use analyses of point data to compare spatial patterns of disturbance and examine multiple drivers of disturbance-generated landscape pattern;
2. compare and contrast the landscape mosaic created by a natural and a human-induced disturbance regime and explore the "historic range of variability" concept; and
3. consider the concept of equilibrium on a disturbance-prone landscape and its sensitivity to spatial and temporal scale.

Students will use a combination of hand calculations, analysis of spreadsheet data, and graphical analyses to understand several different aspects of the interaction between disturbance and landscape pattern. First, students will examine the influence of landscape position on susceptibility to disturbance, using fire initiation patterns in Alaska, USA, as an example. In Part 2, students will examine differences in pattern created by crown fire and harvesting activities in a coniferous landscape in the Rocky Mountains. Finally, students will examine the effects of spatial and temporal scale as they influence inferences about the stability of a landscape.

INTRODUCTION

Disturbance has long been recognized as an important driver of landscape heterogeneity (Watt,. 1947), which is integral to understanding landscape dynamics. A **disturbance** is defined as a relatively discrete event that disrupts the structure of an ecosystem, community, or population and changes resource availability or the physical environment (White and Pickett, 1985). Disturbances are interesting in that they both create and respond to spatial heterogeneity in the landscape, and this is one reason disturbance has received such attention in landscape ecology. Landscape ecological studies of disturbance focus on several aspects of these relationships.

A variety of attributes are used to characterize a disturbance regime. Included among these are the spatial location of the disturbance, the size and shape of disturbed patches, and the spatial extent and frequency of disturbance. In addition, because disturbed landscapes fluctuate through time as succession occurs, the temporal pattern of variation in the landscape is important. Observations of this variation through time in different landscapes have influenced perceptions of how stable a landscape is and whether a particular event results in departures from typical levels of variation.

EXERCISES

Part 1. Landscape Position and Susceptibility to Disturbance

Are various spatial locations in the landscape differentially susceptible to disturbance? If so, can we predict which areas are more or less susceptible to particular types of disturbance? Can locational differences be used to suggest driving factors for particular disturbances? Do the spatial patterns of disturbance locations differ for natural and human-caused disturbances? Susceptibility to disturbance of sites located at particular landscape positions is evaluated by comparing the probability or frequency of occurrence of a particular disturbance at many places in a landscape. A variety of studies have demonstrated that particular locations in a landscape may be more or less sensitive to a given kind of disturbance. For example, Foster (1988a, 1988b) examined a natural disturbance regime characterized by frequent, local events, such as windstorms, pathogens, and lightning strikes, and occasional broad-scale damage by hurricanes and winds in New England. Susceptibility of a site in New

England to disturbance was controlled by slope position and aspect. In landscapes subject to fire, the probability of ignition may vary spatially (e.g., Burgan and Hartford, 1988; Chou et al., 1993). For example, there is a high frequency of lightning ignitions on ridgelines and south-facing slopes in Glacier National Park, Montana (Habeck and Mutch, 1973).

Whereas topographic position does influence fire ignitions, human influences in the landscape may also produce spatial variability in fires. In the Upper Midwestern United States, Cardille et al. (2001) investigated the relationship between wildfire origin locations and environmental and social factors for over 18,000 fires between 1985 and 1995. Results revealed that fires were more likely to occur in locations having higher human population density and less likely to occur in interior forests.

In this exercise, a simple analysis of the spatial distribution of natural and human-caused fire ignitions in Alaska will be used to illustrate the influence of proximity to roads on the spatial pattern of fire and the differences in spatial aggregation of natural and human-caused fires.

Study Site

Natural fires in Alaska occur primarily in the middle third of the state, between the Brooks Range to the north and the Alaska Range to the south (Figure 11.1a). In this interior portion of the state are extensive stands of highly flammable black spruce forests and a continental climate supporting hot, dry summers. Lightning is the major natural source of ignition, with as many as 4000 lightning strikes per day. North of the Brooks Range and along the western coast, lightning strikes are rare and the vegetation is dominated by tundra, whose surface organic mat is normally too wet to burn. South of the Alaska Range, relatively few convective storms generate lightning strikes, and the vegetation is dominated by deciduous forests and deciduous shrub understory, which is also typically too wet to burn. Alaska has relatively few roads, and the majority of the state's population is concentrated in a few towns and cities or dispersed along these roads. Human ignitions account for 85% of the number of fires; however, lightning-caused fires account for more than 90% of the area burned.

EXERCISE 1.1

First, examine the location of natural and human-caused fires in Alaska from the period 1957 to 1975 (Figure 11.1; Gabriel and Tande, 1983). From visual inspection, natural fires seem to be distributed randomly, but the human-caused fires are not. Visually, the human-caused fires appear to be related to the distribution of cities and major roads. To test this relationship, a *t*-test will be used to compare the distances to roads from 20 randomly selected natural fires and 20 randomly selected human-caused fires using a subset of the landscape (Figure 11.2). The hypothesis is that

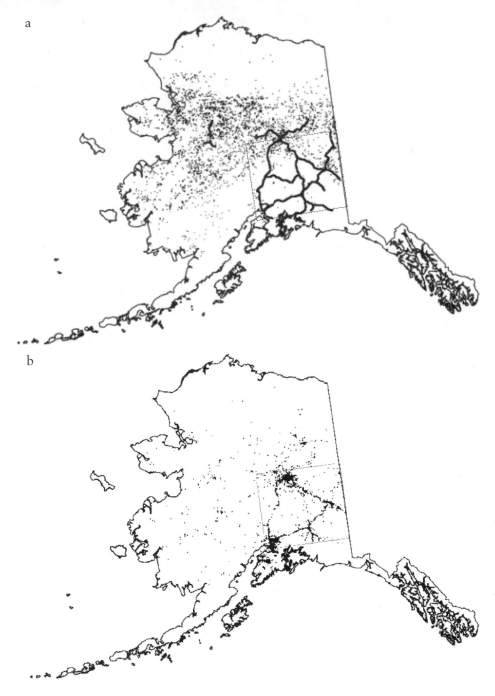

FIGURE 11.1
Maps showing locations of (a) natural fires and (b) human-caused fires in Alaska from 1957 to 1975. The dotted line represents the study landscape used for this exercise and is approximately 250,000 km² in extent (see Figure 11.2a, b). The roads have been removed from Figure 11.1b to better show the locations of human-caused fires.

mean distance to roads will be greater for the natural fires than for those of human origin.

1. Print Figure 11.2a from the CD. Use a calculator or other random number generator to select 20 points of fire initiation from the map of natural fires (Figure 11.2a) as follows:
   (a) For the first randomly selected fire, select a random number between 0 and 135 (the length, in mm, of the x-axis of Figure 11.2a).
   (b) Next, select a random number between 0 and 119 (the length, in mm, of the y-axis of Figures 11.2a and b).
   (c) Plot these two numbers (in mm) as an x–y coordinate on Figure 11.2a and select the fire nearest the x–y coordinate.
   (d) Repeat this until 20 fire locations have been chosen.

a
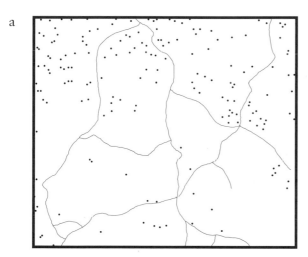

b
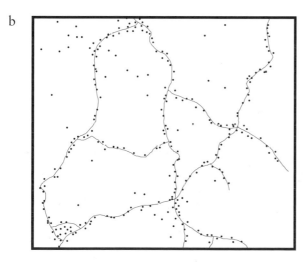

FIGURE 11.2
Maps representing the 250,000-km² study landscapes used for this exercise, showing locations of (a) natural fires and (b) human-caused fires in Alaska. Each point represents a single fire, and solid lines represent roads. Be sure to base your calculations on the printed versions from the CD.

2. Use a ruler to measure the distance (in mm) between each randomly chosen fire and
   (a) the nearest major road
   (b) the nearest fire ("nearest neighbor")

3. Record the distance between each fire and the nearest road in Table 11.1 (which can be printed from the CD), under the column labeled "Distance (mm) on map to nearest road—Natural fire."

4. Record the distance between each fire and its nearest neighbor in Table 11.1, under the column heading "Distance (mm) on map to nearest neighbor—Natural fire."

5. Repeat the analysis (steps 1–3) for the printed map of human-caused fire (Figure 11.1b). Record your results in Table 11.1 under the column "Human-caused fire" for both roads and the nearest neighboring fire.

To test the hypothesis that the mean distance to roads will be greater for natural fires than for human-caused fires, conduct a *t*-test using data from Table 11.1 and Microsoft Excel.

1. Enter the data for the distance to nearest road for both natural and human-caused fires (the first and second columns of data in Table 11.1) into a worksheet in Excel, including the column headings.

2. Under the **Tools** menu, select **Data Analysis.** In the **Analysis Tools** window, scroll down to **"t-test: Two-Sample Assuming Unequal Variances."** (*NOTE:* If Data Analysis does not appear under the **Tools** menu, Select **Add-Ins,** and add the **Analysis Tool Pak.**)

3. Enter the distance to nearest road data from the natural fire and human-caused fire columns into the variable 1 and variable 2 range boxes by highlighting the data in the worksheet using the mouse.

4. Under the **Output Options** box, select the circle next to **Output Range,** and enter a destination cell or range of cells for the *t*-test output in the worksheet, usually a cell a few columns away from your data.

5. Click **OK,** and the *t*-test will be computed.

If the absolute value of the *t*-statistic computed for your data is greater than the *t*-critical value provided in the output, this rejects the null hypothesis, but supports the alternative hypothesis that the mean distance to roads is greater for natural fires than for human-caused fires.

EXERCISE 1.2

To determine whether the spatial dispersion patterns (e.g., random, clumped, or uniform) for natural and human-caused fires are similar or different, you will use a modified Clark and Evans (1954) method of nearest-neighbor analysis. For each fire type, the **expected** mean distance between fires is the density of all fires (number of fires/km^2, or number of fires/mm^2 in map units) within the entire study area. The **observed,** or measured, mean distance between fires is determined using your nearest-neighbor measurements. The ratio of the observed mean distance of fires to the expected mean distance represents a measure of the degree of randomness of the spatial patterns of fire distribution: If the ratio is 1, the fires have a random distribution; if the ratio is 0, the fires are completely aggregated; and a ratio of 2.1419 reflects a completely uniform distribution of the fires.

1. Using either Microsoft Excel or a pocket calculator, calculate the mean distance to the nearest neighbor for natural and human-caused fires (the third and fourth columns of data in Table 11.1). The **expected mean distance between natural fires,** given a density of 180 fires/250,000 km^2, is 18.1 km in the real world, or **4.7 mm** in map units for the purposes of this exercise. The **expected mean distance for human-caused fires,** given a density of 1020 fires/250,000 km^2 is 7.8 km in the real world, or **1.9 mm** in map units for the purposes of this lab.

2. Compute the ratios of the mean observed-to-expected distances for each fire type by dividing the observed value by the expected value.

Question 1.1. Based on your computations, is the spatial pattern of human-caused fires different from the spatial pattern of naturally occurring fires in Alaska? Why or why not?

Question 1.2. In this example, only a single extrinsic variable (distance to roads) was used for the comparison of fire ignition patterns. However, often multiple factors influence the susceptibility of a site to disturbance. If you were to model the ignition of fires across a landscape, what other independent variables might you consider? How might you test for the effects of multiple factors on disturbance initiation and spread? Would these variables be likely to change as the scale of analysis changes from a few hectares to a region, then to continental or global scales? How would they change?

Question 1.3. More generally, under what conditions would you expect landscape position to increase the susceptibility of a site to disturbance? Are there conditions in which landscape position would not be important?

Part 2. Disturbance-Generated Landscape Patterns and the Natural Range of Variability

Disturbances create complex heterogeneous patterns across the landscape because the disturbance may affect some areas but not others, and severity of the disturbance often varies considerably within the affected area. These resulting mosaics may show considerable persistence through time. The spatial patterns created by a variety of natural disturbances have been described (e.g., Foster et al., 1998), and the landscape patterns in human-influenced landscapes have been compared with those subjected to natural disturbances only (e.g., Krummel et al., 1987; Mladenoff et al., 1993). In particular, the landscape patterns resulting from forest harvesting strategies have received considerable attention (e.g., Franklin and Forman, 1987; Li et al., 1993; Gustafson and Crow, 1996), and a number of comparative studies have examined the differences in the landscape mosaic resulting from wildfire and forest harvesting. For example, Delong and Tanner (1996) compared the spatial characteristics of landscapes in British Columbia subjected to regularly dispersed 60- to 100-hectare clearcuts with the historic patterns generated by wildfire. They found that wildfires created a more complex landscape mosaic that included a greater range of patch sizes and more complex disturbance boundaries. In addition, individual wildfires were often greater than 500 hectares in size, but unburned forest patches remained within the perimeters of the fire (Delong and Tanner, 1996).

There has been considerable discussion about the use of natural spatial patterns as a model for the pattern and timing of human disturbances (e.g., clearcutting) (Hunter, 1993; Attiwill, 1994; Holling and Meffe, 1996), with the implicit assumption that ecological processes will be better maintained in this way. For example, Runkle (1991) suggested that temperate deciduous forest should be harvested in a pattern that mimics small treefall gaps, whereas Hunter (1993) recognized that boreal forests would require very large clearcuts if they were to imitate the size and arrangement of boreal fires. Improved understanding is needed of the nature and dynamics of disturbance-generated mosaics in a wide variety of landscapes and how these differ from human-generated patterns. Although this idea is intuitively appealing, it is difficult to define and mimic natural patterns objectively. Meeting such an objective also requires understanding the dynamics of the natural disturbance regime in a given landscape and the range of variation to be expected in its spatial pattern.

The use of historical patterns and processes as reference conditions for informed land management has emerged as an increasingly recognized and debated concept in ecosystem management (Parsons et al., 1999). **Natural variability** is defined as the spatial and temporal variation in ecological conditions that are relatively unaffected by people within a period of time and geographic area appropriate to an expressed goal (Landres et al., 1999). The concepts considered under the rubric of "range of natural variation" or "natural variability" include (1) that disturbance-driven spatial and temporal variability is a vital attribute of nearly all ecological systems, and (2) that past conditions and processes provide context and guidance for managing ecological systems today (Landres et al., 1999). Using these concepts in ecosystem management requires understanding the history of a given landscape and knowing its disturbance regime.

This exercise will involve comparing the differences in landscape pattern that are produced by two different disturbances—naturally occurring wildfire and forest harvesting—and assessing whether the forest harvesting moves the landscape mosaic out of a 300-yr range of natural variation. Both landscapes are coniferous forests dominated by lodgepole pine and located in the Rocky

Mountain region in Wyoming (USA). Large, infrequent crown fires are the dominant natural disturbance (Turner and Romme, 1994), but the region has been subject to intensive timber management since the early part of the 20th century.

Study Site

The reference landscape is a 14.7 × 16.3-kilometer section of Yellowstone National Park (YNP). The landscape mosaic for the past 250 years has been reconstructed based on dendrochronology and fire history at approximately 100-year intervals based on Tinker et al. (in review) (Figure 11.3). The human-influenced landscape is a 14.7 × 16.3-kilometer section of the Medicine

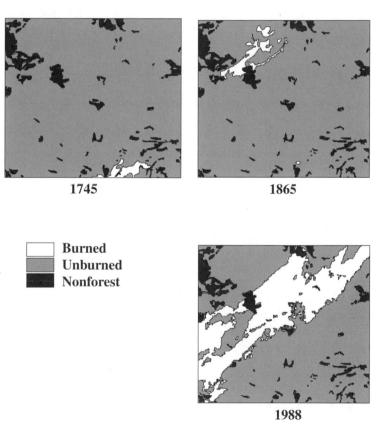

FIGURE 11.3
Digital maps of three of the nine landscape mosaics for Yellowstone National Park, Wyoming, used in the FRAGSTATS analyses of landscape metrics. All maps depict the same ~240-km² region. The 1745 map represents a period of few fires; the 1765 map represents a period during which a moderate amount of the landscape was burned; and the 1988 map represents the landscape following the widespread, intense fires of 1988.

Bow National Forest (MBNF), which is similar to the YNP landscape but has undergone extensive forest harvesting (Figure 11.4). Two alternative cutting patterns are depicted: (1) the historical harvest (Figure 11.4a and b), in which large clearcuts were separated by long, narrow buffer strips of uncut forest, and (2) the more recent harvest pattern (Figure 11.4c and d), designed to better reflect natural patterns.

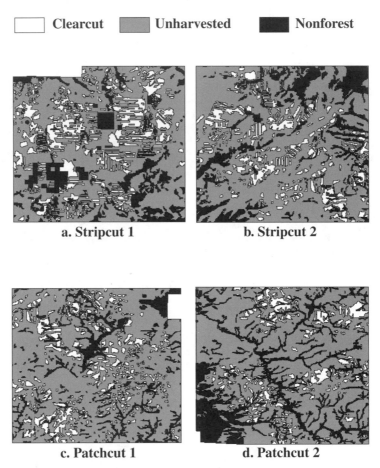

FIGURE 11.4
Digital maps of the Medicine Bow National Forest, Wyoming, used in the FRAGSTATS analyses of landscape metrics. All maps depict the same ~240-km² region. The upper two maps, stripcut 1(a) and stripcut 2(b), represent clear-cut harvesting patterns practiced during the 1960s and 1970s, when long, linear patches of forest were harvested, leaving narrow strips of uncut forest ~50–100 m wide between. The lower two maps, patchcut 1(c) and patchcut 2(d), represent clearcut harvesting patterns practiced during the 1980s and 1990s. The shape and allocation of these clearcuts have been designed to better reflect natural patterns.

EXERCISE 2.1

Spatial analyses were conducted on each of these study landscapes using Fragstats (McGarigal and Marks, 1995). The output from the spatial analyses can be found in a spreadsheet data file (**cut_burn.xls**) on the CD, under the directory for this lab.

1. After the **cut_burn.xls** file has been opened, use the **Save As** command in Excel to copy the file to a location on the hard drive or a floppy disc. This is done so that modifications to the file may be made.

2. Under **Tools** and **Data Analysis** in Excel, calculate **Descriptive Statistics** for each of the YNP landscape metrics—use the **Fire-generated landscape** section of the spreadsheet. After selecting a column of data to analyze (the input range), you must also choose an output range and be sure that "Summary statistics" is checked. For example, generate descriptive statistics for the number of burned patches in YNP from 1705 to 1988. Among the statistics generated by Excel will be (a) the mean, and (b) the standard deviation for each column of data.

3. Calculate the coefficient of variation (CV) for each of the YNP metrics using the formula:

$$CV = 100 \times \text{standard deviation} / \text{mean}$$

The CV is a relative measure of the variation in the data, independent of the unit of measure of the original data, expressed as a percentage of the sample mean. Compare the mean and CV values for each metric in the YNP landscape ($n = 9$) to the corresponding values for each metric for the MBNF landscape ($n = 4$) listed in the cut_burn.xls file.

4. Using the functions within Excel, generate the following plots (metric vs. year) for YNP:
   (a) Number of patches
   (b) Mean patch size
   (c) Percent of landscape (occupied by each patch type)
   (d) Edge density (edge length in meters per unit area of the landscape)
   (e) Total core area index (percentage of burned or unburned core area in the landscape)
   Plot burned and unburned patches separately (total of 9 plots − total core area index data only for unburned patches). For example, use the nine map years (e.g., 1705, 1745, etc.) as the x-axis values, and the number of patches for the y-axis values. Label each axis and provide a title describing the metric that was plotted.

5. After plotting the YNP data on each graph, double-click on the **Y-axis**—this will show the **Format Axis** dialog box. Click on the **Scale**

tab and ensure that the **Maximum Value (Y) Axis Scale** is at least as large as the largest corresponding value from the MBNF data. This is necessary because the MBNF data will be added to the graph by hand in a later step. You may need to change the minimum value as well to accomodate the MBNF data.

6. Print out hard copies of each of the nine plots of YNP Metrics. Draw two dotted lines on each graph. One line should pass through the maximum value and the other line should pass through the minimum value. The range of values **between** the two dotted lines may be thought to represent an historic range of variability, or reference landscape for naturally developing forests in the central Rocky Mountains.

7. Using a pencil, add data points representing corresponding values for the four MBNF landscapes (patchcuts 1 and 2; stripcuts 1 and 2) to the appropriate YNP plot. Add clearcut values to plots containing burned values from YNP, and add uncut values to plots containing unburned values from YNP. These plots will be used to answer the questions below.

Question 2.1. Describe the range of natural variability for the spatial pattern of forests in the YNP (reference) landscape. Use the full set of metrics and their variability (CV as well as the plots showing their temporal trends) to develop your description.

Question 2.2. Describe the spatial patterns of forests in the National Forest (managed) landscape. Again, use the full set of metrics that were computed.

Question 2.3. Compare the spatial patterns in the reference and managed landscapes. Which aspects of spatial pattern are similar? Which are different?

Question 2.4. Based on the quantitative assessment, does the landscape produced by the *historic* pattern of clearcutting fall within the range of natural variability of the reference landscape?

Question 2.5. Based on the quantitative assessment, does the landscape produced by the *recent* pattern of harvesting fall within the range of natural variability of the reference landscape?

Question 2.6. Do your interpretations of similarity or difference between reference and managed landscapes change depending on what metric you consider? What recommendations might you make to managers regarding how many metrics and which ones should be considered when evaluating dynamics of a disturbance-prone landscape?

Question 2.7. If a goal of landscape management is to have human disturbances mimic natural disturbances, what factors must be considered in addition to the spatial pattern and timing of the disturbance events?

Part 3. Disturbance and Landscape Equilibrium

The natural range of variability concepts emerged from the recognition that landscapes are constantly changing, and that many ecological systems cannot be described as equilibrial (Wu and Loucks, 1995). Watt (1947) first proposed the notion of a dynamic spatial mosaic that produces a stable distribution of successional stages at the landscape level. Over the years, questions of whether equilibrium can be detected on landscapes subject to disturbance, and how large a landscape must be to incorporate a given disturbance regimes, have been important themes in landscape ecology (e.g., Shugart and West, 1981; Romme, 1982; Baker, 1989; Turner and Romme, 1994). Much of the disagreement surrounding equilibrium versus nonequilibrium, and stability versus instability, can be attributed to several factors: the ambiguity in various definitions, different views of spatial heterogeneity and its effects, the lack of explicit specification of scales, and differences in theoretical foundations (Turner et al., 1993). Landscapes can exhibit a variety of behaviors under different disturbance regimes, and the same landscape may shift among different regions of behavior.

This exercise examines the temporal dynamics that may be observed in a landscape by using a simplified version of the model used by Turner et al. (1993) to explore landscape equilibrium. The landscape will be assumed to contain three potential seral stages: a pioneer, mid-, and late-successional class. A disturbance that affects a grid cell will return that cell to the pioneer stage, which persists until the next time step. The midsuccessional class also persists for one time step. The late-successional stage persists until a disturbance event resets it to the pioneer stage.

The landscape is represented as a grid of cells. Disturbance events occur at a prescribed frequency and size. Two integrated parameters relate the size of the disturbance to the size of the landscape, and the frequency of the disturbance to the recovery time of the vegetation. The **spatial parameter** S equals the disturbance size/landscape size. The parameter S varies between 0 and 1 and is easily interpreted as the proportion of the landscape disturbed per event. For example, if a disturbance affects 10 hectares of a 100-hectare landscape, then $S = 0.10$. In this exercise, the disturbance size and the landscape size will both be varied. The **temporal parameter,** T, equals disturbance return interval/recovery time. If disturbances occur every ten years and a disturbed site recovers to the mature successional stage in five years, then $T = 10/5 = 2$. The parameter T has three qualitatively different states that are useful to understand. When $T = 1$, then the disturbance interval is equal to the recovery time, and there is adequate time for a disturbed cell to reach the mature stage before again being disturbed. When $T < 1$, the disturbance will recur before full recovery has been achieved. Thus, if the disturbance is also large, the landscape can be effectively maintained in an early pioneer stage. When $T > 1$, the disturbed sites recover fully and persist in the mature state for some time before the next disturbance event occurs. In this exercise, the recovery time will always be set to three time steps, but the interval between disturbances will be varied.

EXERCISE 3.1

Consider how the model operates by first describing the changes that would occur in one cell (thus $S = 1.0$), just to assure that the basic operation of the model is clear. Examine and print out Figure 11.5 on the CD, which provides appropriately sized landscape grids for use in the next several exercises.

1. Consider a disturbance that occurs every 5 years (i.e., $T = 5/3 = 1.67$). Enter a 3 in the first cell in Figure 11.5a, then have the disturbance set the cell back to a 1 during the next time step (each cell represents the landscape at a different time step). Continue writing the number in the cell at each time step to describe the successional stage of the cell (1 = pioneer, 2 = mid, 3 = late).

2. Using either a pencil and paper or Excel, plot the proportion of the landscape occupied by each successional stage at each of the 12 time steps. The y-axis will be scaled from 0 to 1, and the x-axis will extend from 1 to 12. The plot will contain three lines, one for each successional stage.

Question 3.1. Next, assume that the disturbance occurs each year. What would happen to the relative abundances of successional stages present on that landscape? Which ones would be lost, and why?

EXERCISE 3.2

Next, you consider disturbances of two different sizes: one cell (in Exercise 3.2) and four cells (in Exercise 3.3) on a landscape that contains nine cells. The return interval of the disturbance is set to 1 year ($T = 0.33$).

1. On Figure 11.5b, begin by penciling a 3 in each of the open grid cells of the 3×3 landscape to represent the initial landscape.

2. For the second time interval, simulate a one-cell disturbance. Select a cell to be disturbed at random, and reset that cell to the pioneer stage by entering a 1. As the remaining cells were not disturbed, enter a 3 in each remaining cell.

3. At the third time step, another cell is disturbed at random. Choose a cell and enter a 1 in that cell. The pioneer cell that was disturbed at the last time step makes the transition to midsuccessional stage, so record a 2 in the cell. Note that the same cell might be disturbed in two successive time steps; if that occurs, then the cell will remain at stage 1.

4. Continue with this procedure, penciling in the successional stages on the landscape for each of the 12 time steps, assuming a one-cell disturbance.

5. Compute the proportion of the landscape occupied by each successional stage at each time step and plot this; you will have one graph
with three lines (one for each stage).

EXERCISE 3.3

Now, the size of the disturbance will be increased to four cells (a 2×2
square), and the disturbance will still occur each year.

1. Repeat the procedure that you used in Exercise 3.2, keeping the disturbance frequency the same but imposing the larger disturbance on
the landscape (use Figure 11.5c).

2. As you did earlier, plot the proportion of the landscape occupied by
each successional stage under each of these two scenarios for 12 time
steps.

3. Compute the spatial parameter, *S*, for each of the disturbance sizes
in Exercises 3.2 and 3.3.

Question 3.2. Compare the plots for the two landscapes in Exercises 3.2 and
3.3. How do the landscapes differ in terms of their relative abundances of the
pioneer and mature successional stages? Do the landscapes show similar or
different levels of fluctuation through time?

EXERCISE 3.4

1. Repeat the preceding procedure for the four-cell disturbance but
change the disturbance return interval to 5 years (Figure 11.5d). Compute the proportions occupied by each successional stage through
time, and plot these for each of the 12 time steps.

2. Compute the spatial parameter, *S*, and temporal parameter, *T*.

Question 3.3. When the return interval of the disturbance was increased, did
the fluctuations in the proportions of the landscape occupied by each successional stage change? Do you interpret this as moving toward more or less stable conditions? How do you weight the relative proportions of the successional stages and their relative fluctuations?

Question 3.4. If the spatial extent of the landscape was increased to 36 cells
(a 6×6 grid, see Figure 11.5e), what would happen to the changes through
time in the proportions of successional stages 1, 2, and 3? (*NOTE:* you can
compute this easily by changing the denominator that you used to compute
the proportions in the preceding example from 9 to 36 for stages 1 and 2;
note that the number of cells for stage 3 will fluctuate). If the landscape was
reduced in size to four cells, what would happen to the changes through time?
Does the landscape appear more or less stable with these changes in spatial
extent?

Results from many computer-based simulations using a similar simple model were used to generate a state space describing the variation in a land-scape for disturbance regimes characterized by different values of T and S (Turner et al., 1993, Figure 11.6). When disturbances are small relative to the landscape, and the return interval is long relative to the recovery time, the landscape will be dominated by the mature successional stage and will fluc-tuate little through time. When the disturbances are very large relative to the size of the landscape, the proportions of each successional stage will fluctu-ate widely on the landscape.

Question 3.5. From your calculations and from considering the state-space diagram, describe the perceived stability of a landscape under the following disturbance regimes:
  (a) Small, frequent disturbances
  (b) Small, infrequent disturbances
  (c) Large, frequent disturbances
  (d) Large, infrequent disturbances

Question 3.6. What are the advantages and disadvantages of using such a simple model to represent a disturbance regime on a landscape? Should other factors also be included in such a conceptualization? Why or why not?

Question 3.7. How does the state space defined by S and T relate to the con-cept of the natural range of variability addressed in Part 2? Can this concep-tualization aid landscape managers? Why or why not?

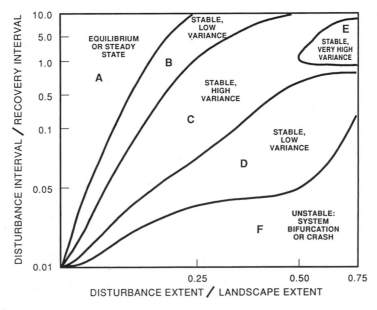

FIGURE 11.6
Landscape disturbance state space. Adapted from Turner et al. (1993).

CONCLUSIONS

The preceding exercises have demonstrated how the initiation of disturbance can be affected by landscape position, how disturbances create pattern and dynamics on landscapes, and how spatial and temporal scale influences whether a landscape is equilibrial. You've also seen the wide range of variability in natural disturbance regimes and the difficulty in setting reference conditions for a disturbance. Either explicitly or implicitly, each of these examples also illustrates the potential importance of humans in governing disturbance dynamics. The first example showed how the use of the landscape for purposes seemingly unrelated to the disturbance in question can influence the pattern of disturbance (e.g., roads influencing the initiation of fires). Direct use of the landscape for forest harvest can also substantially alter landscape patterns. In the last example, the parameters S and T were used to estimate how human-induced changes in a disturbance regime (e.g., alteration of disturbance size or frequency) might push the landscape into regions of different behavior. Humans might alter the size of the disturbance relative to the size of the landscape (S) by either altering the extent of fire or by altering the extent of the landscape itself (through land conversion and fragmentation). Furthermore, humans could alter landscape disturbance by burning or cutting areas larger than might otherwise be disturbed. The recovery time of the disturbance relative to the recovery time of the vegetation (T) can also be altered by humans—by initiating disturbance more frequently through altering the frequency of fires or frequency of harvests. Thus, humans can influence the spatial and temporal dynamics of disturbance, creating another layer of complexity in addition to the natural variability in disturbance dynamics, which should not be ignored.

BIBLIOGRAPHY

Note. An asterisk preceding the entry indicates that it is a suggested reading.

ATTIWILL, P. M. 1994. The disturbance of forest ecosystems: The ecological basis for conservative management. *Forest Ecology and Management* 63:247–300.

BAKER, W. L. 1989. Landscape ecology and nature reserve design in the Boundary Waters Canoe Area, Minnesota. *Ecology* 70:23–35.

*BOOSE, E. R., D. R. FOSTER, AND M. FLUET. 1994. Hurricane impacts to tropical and temperate forest landscapes. *Ecological Monographs* 64:369–400. Uses empirical data and modeling to examine the sensitivity of different landscape positions to hurricane effects in New England (USA) and Puerto Rico.

BURGAN, R. E., AND R. A. HARTFORD. 1988. Computer mapping of fire danger and fire locations in the continental United States. *Journal of Forestry* 86:25–30.

CARDILLE, J. A., S. J. VENTURA, AND M. G. TURNER. 2001. Environmental and social factors influencing wildfires in the Upper Midwest, USA. *Ecological Applications* 11(1):111–127.

CHOU, Y. H., R. A. MINNICH, AND R. A. CHASE. 1993. Mapping probability of fire occurrence in San Jacinto Mountains, California, USA. *Environmental Management* 17:129–140.

CLARK, P. J., AND F. C. EVANS. 1954. Distance to nearest neighbor as a measure of spatial relationships in populations. *Ecology* 35(4): 445–453.

DALE, V. H., A. E. LUGO, J. MACMAHON, AND S. T. A PICKETT. 1998. Ecosystem management in the context of large disturbances. *Ecosystems* 1:546–557.

DELONG, S. C., AND D. TANNER. 1996. Managing the pattern of forest harvest: Lessons from wildfire. *Biodiversity and Conservation* 5:1191–1205.

FOSTER, D. R. 1988a. Disturbance history, community organization and vegetation dynamics of the old-growth Pisgah Forest, southwestern New Hampshire, USA. *Journal of Ecology* 76:105–134.

FOSTER, D. R. 1988b. Species and stand response to catastrophic wind in central New England, USA. *Journal of Ecology* 76:135–151.

*FOSTER, D. R., D. H. KNIGHT, AND J. F. FRANKLIN. 1998. Landscape patterns and legacies resulting from large infrequent forest disturbances. *Ecosystems* 1:497–510. Focuses on disturbances in forest landscapes and provides a nice comparison of the variety of spatial patterns introduced by large, infrequent disturbances and their persistent effects on the landscape.

FRANKLIN, J. F., AND R. T. T. FORMAN. 1987. Creating landscape patterns by forest cutting: Ecological consequences and principles. *Landscape Ecology* 1:5–18.

GABRIEL, H. W., AND G. F. TANDE. 1983. A regional approach to fire history in Alaska. BLM-Alaska Technical Report 9. U.S. Bureau of Land Management, Anchorage.

GUSTAFSON, E. J., AND T. R. CROW. 1996. Simulating the effects of alternative forest management strategies on landscape structure. *Journal of Environmental Management* 46:77–94.

HABECK, J. R., AND R. W. MUTCH. 1973. Fire-dependent forests in the northern Rockies. *Quaternary Research* 3:408–424.

HOLLING, C. S, AND G. K. MEFFE. 1996. Command and control and the pathology of natural resource management. *Conservation Biology* 10: 328–337.

HUNTER, M. L. JR. 1993. Natural fire regimes as spatial models for managing boreal forests. *Biological Conservation* 65:115–120.

KRUMMEL, J. R., R. H. GARDNER, G. SUGIHARA, R. V. O'NEILL, AND P. R. COLEMAN. 1987. Landscape patterns in a disturbed environment. *Oikos* 48:321–324.

*LANDRES, P. B., P. MORGAN, AND F. J. SWANSON. 1999. Overview of the use of natural variability concepts in managing ecological systems. *Ecological Applications* 9:1179–1188. This paper is part of a Special Feature of Ecological Applications and provides an excellent overview of the concept of historic range of variation; interested readers are encouraged also to read the other papers published in this Special Feature.

LI, H., J. F. FRANKLIN, F. J. SWANSON, AND T. A. SPIES. 1993. Developing alternative forest cutting patterns: A simulation approach. *Landscape Ecology* 8:63–75.

MCGARIGAL, K., AND B. J. MARKS. 1995. FRAGSTATS. Spatial analysis program for quantifying landscape structure. USDA Forest Service General Technical Report PNW-GTR-351.

MLADENOFF, D. J., M. A. WHITE, J. PASTOR, AND T. R. CROW. 1993. Comparing spatial pattern in unaltered old-growth and disturbed forest landscapes for biodiversity design and management. *Ecological Applications* 3:293–305.

PARSONS, D. J., T. W. SWETNAM, AND N. L. CHRISTENSEN. 1999. Uses and limitations of historical variability concepts in managing ecosystems. *Ecological Applications* 9:1177–1178.

PICKETT, S. T. A., AND P. S. WHITE, EDS. 1985. *The Ecology of Natural Disturbance and Patch Dynamics*. Academic Press, New York.

RISSER, P. G., J. R. KARR, AND R. T. T. FORMAN. 1984. Landscape ecology: Directions and approaches. Special Pub. No. 2. Illinois Natural History Survey, Champaign.

ROMME, W. H. 1982. Fire and landscape diversity in subalpine forests of Yellowstone National Park. *Ecological Monographs* 52:199–221.

ROMME, W. H., AND D. H. KNIGHT. 1982. Landscape diversity: The concept applied to Yellowstone Park. *BioScience* 32:664–670.

RUNKLE, J. R. 1991. Gap dynamics of old-growth eastern forests: Management implications. *Natural Areas Journal* 11:19–25.

SHUGART, H. H. JR., AND D. C. WEST. 1981. Long-term dynamics of forest ecosystems. *American Scientist* 69:647–652.

TINKER, D. B., W. H. ROMME, AND D. G. DESPAIN. Historic range of variability in landscape structure in subalpine forests of the Greater Yellowstone Area. Manuscript in review.

TURNER, M. G., ed. 1987. *Landscape Heterogeneity and Disturbance.* Springer-Verlag, New York.

TURNER, M. G., AND V. H. DALE. 1998. Comparing large, infrequent disturbances: What have we learned? Introduction for special feature. *Ecosystems* 1:493–496.

TURNER, M. G., V. H. DALE, AND E. E. EVERHAM III. 1997. Fires, hurricanes and volcanoes: Comparing large-scale disturbances. *BioScience* 47:758–768.

TURNER, M. G., R. H. GARDNER, V. H. DALE, AND R. V. O'NEILL. 1989. Predicting the spread of disturbance across heterogeneous landscapes. *Oikos* 55:121–129.

TURNER, M. G., AND W. H. ROMME. 1994. Landscape dynamics in crown fire ecosystems. *Landscape Ecology* 9:59–77.

TURNER, M. G., W. H. ROMME, R. H. GARDNER, R. V. O'NEILL, AND T. K. KRATZ. 1993. A revised concept of landscape equilibrium: Disturbance and stability on scaled landscapes. *Landscape Ecology* 8:213–227.

WATT, A. S. 1947. Pattern and process in the plant community. *Journal of Ecology* 35:1–12.

WHITE, P. S., AND S. T. A. PICKETT. 1985. Natural disturbance and patch dynamics: An introduction. In S. T. A. Pickett and P. S. White, eds. *The Ecology of Natural Disturbance and Patch Dynamics.* Academic Press, New York, pp. 3–13. This chapter does an excellent job of defining disturbance and its various attributes. The rest of the book provides good examples of disturbance from a variety of systems.

*WU, J., AND O. L. LOUCKS. 1995. From balance of nature to hierarchical patch dynamics: A paradigm shift in ecology. *Quarterly Review of Biology* 70:439–466. Provides a comprehensive and thoughtful treatment of how patch dynamics are considered across a wide range of scales. This is a "must read" for those interested in scale-dependent dynamics and the effects of disturbance on landscapes.

ALTERNATIVE STABLE STATES

GARRY D. PETERSON

OBJECTIVES

Ecological structure and process can interact to produce and maintain specific ecological organizations. Ecological organizations that encourage their own persistence are often described as ecological stable states, because when disturbance alters these systems, the interaction of process and structure tends to move it back toward that state. Many ecosystems, including coral reefs, rangelands, and shallow lakes, appear to possess more than one stable or self-maintaining state (Dublin et al., 1990; Knowlton, 1992; Scheffer et al., 1993). This lab will examine alternative stable states and their impact on landscape dynamics in order to help you

1. understand how alternative stable states can emerge in ecosystems;
2. understand how change can be analyzed using a transition matrix;
3. understand how spatial interactions can alter a site's dynamics; and
4. examine the resilience and brittleness of landscape patterns.

In this lab, you will examine alternative stable states using two related models of north Florida forest fire dynamics. The first model explores how ecological interactions can produce alternative stable states in ecosystems and shows how these dynamics can be analyzed. The second model explores how landscape pattern can influence the spread of disturbance and maintain or eliminate alternative stable states.

The models are contained in two Excel spreadsheets. The first model is simpler than the second; it is a nonspatial site model (**site.xls**) that represents forest fire dynamics in a small patch of forest. The second spreadsheet (**land.xls**) contains a spatial model that represents an entire landscape by connecting a set of the site models from the first spreadsheet and allowing fire to spread among sites.

INTRODUCTION

Part 1. Alternative Stable States in a Forest Savannah

Much of the southeastern United States used to be covered by longleaf pine forest (Rebertus et al., 1989). However, in the past centuries, human activities such as logging, agriculture, and fire suppression have reduced the area covered by longleaf pine forest to less than 5% of its former range. One of the largest remaining areas of longleaf pine remaining is located in northeastern Florida.

Either longleaf pines (*Pinus palustris*) or hardwoods such as oaks (*Quercus* spp.) can dominate the vegetation of the sandy soils of northeastern Florida (Glitzenstein et al., 1995). The ground vegetation in these forests burns frequently, and because longleaf pine and oaks have different responses to and effects on fire, fire mediates the competitive relationships between these two vegetation types. Both young and mature longleaf pines can survive ground fires. Additionally, mature longleaf pines shed needles that provide good fuel for ground fires. Young oaks, however, are intolerant of fire, but mature oaks shed leaves that suppress the buildup of good fuel for ground fires, encouraging the growth of young oaks. Without fire, oaks can recruit in and eventually replace longleaf pine stands. However, regular fires maintain longleaf pine by suppressing oak growth, resulting in more fuel accumulation in pine stands, which encourages more fire, further suppressing hardwoods.

Site Model

A simplified version of the ecological dynamics of northern Florida forests are represented in the site model, contained in the file **site.xls** on the CD. To make this model tractable in a spreadsheet, the relationship between fire and vegetation has been simplified, and the speed with which longleaf pine can be converted to hardwood has been accelerated. However, despite these changes, the model captures the fundamental dynamics of the interactions between oaks, longleaf pine, and fire in northern Florida.

This model represents the dynamics of a small, 50×50-meter, patch of forest. The model contains five variables. Three of these variables determine the rate at which a site's probability of combustion declines in the absence of fire. The other two variables determine a site's fire ignition rate and the initial time since the site was last burned. The model outputs the temporal dynamics of a site's fire history, and consequently its vegetation. The model assumes that sites that have recently burned are dominated by longleaf pine, and that sites that do not burn are increasingly dominated by hardwood, until after 25 years hardwood is the dominant vegetation. The **Time Since Fire** (**TSF**) is the central variable that is used to describe the dynamics of a site in this model.

You should open the spreadsheet site.xls and examine these parameters. When opening the file ignore any messages related to links to other worksheets. The spreadsheet is an Excel workbook that contains four separate pages. Tabs at the lower left of the screen indicate these pages. Make sure that you begin on the page "Site Model."

Model Input

Two main parameters control the rate of fire initiation (P_{ignit}) and the initial condition of the model (TSF_0). These parameters are colored light green on the spreadsheet.

Probability of Fire Ignition (cell F4) determines the rate of fire ignition measured in ignitions per year. In this model there can be no more than one fire ignition per year, and consequently the Probability of Fire Ignition must be between 0 and 1. This variable will be abbreviated as P_{ignit}.

Initial Time Since Fire (cell F7) sets the initial vegetation of the site by specifying how long it has been since the site last burned. The Time Since Fire can range between 1 and 25 years. While a site may not burn for over 25 years, the changes in vegetation that occur after 25 years have minimal impacts on any subsequent fires, thus, these dynamics are not explicitly represented in this model. Initial Time Since Fire is the abbreviated as TSF_0, because it refers to the Time Since Fire at time zero of the model run.

A second set of input parameters determines the probability that a site will burn if it is ignited, or the probability of combustion. These parameters represent the forest ecology of northern Florida (Peterson, 1999) and will not vary as much as the first set of parameters.

The **Probability of Combustion function** determines how the probability of a site catching fire changes with its Time Since Fire (*TSF*). This probability function decays as the Time Since Fire increases. This decline represents the gradual accumulation of fire-inhibiting hardwood species, such as oak, in the site and is represented by a logistic function.

Three variables define this logistic function and are colored light blue in the spreadsheet.

K (cell C6)—the maximum theoretical probability of burning

P_0 (cell C4)—the decline in the probability of combustion from *K* at time zero. This sets the value of the probability of combustion function at time zero.

R (cell C5)—the rate at which probability of combustion declines over time

Mathematically, P_{comb} (the probability of combustion) is defined as K minus a logistic function of *TSF* defined by K, P_0, and R:

$$P_{Comb}(TSF) - K - \cfrac{R}{\dfrac{R}{K} + \left(\dfrac{R}{P_0} - \dfrac{R}{K}\right)e^{-R \times TSF}}$$

Examine cell C10 and confirm that this function is being used.

On the left of the spreadsheet are two graphs, Figures 12.1 and 12.2. The top left graph (at cell D9), **Figure 12.1, Probability of Combustion,** illustrates the function $P_{comb}(TSF)$ that is defined by the P_0, R, & K input parameters [cells C4–C6]. This graph represents the decay in the combustibility of a site with increasing Time Since Fire, as the accumulation of oak litter inhibits fire.

At the lower left-hand side of the spreadsheet, **Figure 12.2, Probability of Fire or No Fire,** plots the P_{fire} function (the probability that a site that has not burned for a specific number of years burns during a year) and P_{nofire} (the probability that a site does not burn). P_{fire} is the product of the probability of a site being ignited and the probability of fire spread ($P_{fire} = P_{ignit} \times P_{comb}$). The probability of a site not burning is defined as $P_{nofire} = 1 - P_{fire}$.] Since P_{comb} is a function of *TSF*, both P_{fire} and P_{nofire} are also functions of *TSF*.

Model Output

The graphs on the right of the spreadsheet represent the output of a stochastic model configured by the input variables. **Figure 12.3, Site Dynamics of TSF** (at cell I3), displays change in *TSF* over time. **Figure 12.3** shows the model beginning at the initial conditions and then changing following the dynamics defined by the input parameters. **Figure 12.4, Occurrence of Fire** (at cell I15), illustrates the fire history of the site. The state of the model changes every year, and the dynamics of the model are stochastic. *TSF* increases one year if there is no fire. If there is a fire, *TSF* is set back to zero. The state of the model after 100 years of simulation, TSF_{100}, is shown by the red number in the yellow box (cell K24).

Instructions for Running the Site Model

Set the fire spread function parameters to P_0 (cell C4): 0.00125; R (cell C5): 0.65; K (cell C6): 0.975; and P_{ignit} [cell F4] to 0.45 and TSF_0 [cell F7] to 1.

On a Windows machine F9 will recalculate the model, while on a Macintosh, pressing **Command + =** will recalculate the model. (*NOTE:* Some Windows versions of Excel may work differently. If **F9** fails to recalculate the model, you should learn how to recalculate a spreadsheet using Excel's help menu.) Recalculate the model several times and observe the model's variable behavior.

How Does the Model Work?

Now that you have looked at the model's output, you should see how those results are produced.

1. Move the spreadsheet window to the right and select cell AI3. This cell contains the initial conditions of the model (TSF_0 copied from cell F7).

2. Move down to cell AI5, and note that it calculates the probability of combustion of a site having the TSF of cell AI3 using the R, K, and P_0 values.

3. Go down to AI6. This cell multiplies the P_{comb} (calculated in AI5) with the parameter P_{ignit} (from cell F4). This value is P_{fire}, the overall probability that a site will catch fire this year.

4. Continue downward to AI9. This cell contains a logical statement that contains the stochastic portion of the model. A pseudorandom number between 0 and 1 is calculated. If this number is greater than P_{fire}, then the site burns; otherwise, it does not burn. One represents the site burning, and zero represents it not burning.

5. Recalculate the model several times. You should notice how the value of cell AI9 changes from 1 to 0.

Cell AI10 uses the outcome of the stochastic simulation in cell AI9 to calculate the **Time Since Fire** for the following year. If the model did burn, then TSF is set back to 1, because in the following year it is 1 year after a fire. If the model did not burn, TSF is increased by 1. You should note that the **TSF** is not allowed to exceed 25. This constraint simplifies the model by assuming there is no functional difference between sites that have gone 25 years without fire and those that remained unburned for longer periods.

Next, move to the top of the spreadsheet and over to the next column to cell AJ3. In this cell the Time Since Fire of the model is set to the value calculated in cell AI10. This updated value is then updated itself in the column AJ. Repeating this process 100 times produces the model's behavior. The row of TSF values, starting at cell AI3, is displayed in **Figure 12.3**. The row of fires, starting at cell AI9, is displayed in **Figure 12.4**. Recalculate the model and note how the calculated values of TSF change.

EXERCISES

EXERCISE 1.1

Return to top left of the spreadsheet (cell A1). Your first manipulation of the model will be to vary some model parameters.

1. Set the initial values in the model to $P_0 = 0.00125$, $R = 0.65$, $K = 0.975$. Observe the graph that these settings produce in Figure 12.1, Probability of Combustion, in the spreadsheet.

2. Alter P_0, R, and K and observe how this function changes. Decrease K to 0.9, 0.8, 0.7, 0.5, and 0.2. Increase R to 5, and then decrease R to 3, 1, 0.5, 0.25, 0.1. Set P_0 to 0.001, 0.005, 0.01, 0.05, and 0.1. Set each value back to original before changing the next.

3. Set the initial values in the model to the initial values $P_0 = 0.00125$, $R = 0.65$, $K = 0.975$. This fire spread function represents the inhibition of combustion that occurs as oaks accumulate. Answer Question 1.1 below.

4. Try increasing and decreasing P_{ignit}. Set P_{ignit} to 1, which means there is one fire every year, and compare the shape of the P_{fire} function to that of the P_{comb} function. Now decrease P_{ignit} to 0.5. Answer Question 1.2 below.

Question 1.1. If oaks grew slower and more slowly decreased the P_{comb}, then how would you alter R, K, and/or P_0 to represent this change?

Question 1.2. How does a higher P_{ignit} value alter the relationship between the P_{comb} function and the P_{fire} function, compared to a lower P_{ignit} value?

EXERCISE 1.2

1. Change P_{ignit} back to 0.45. Manipulate the model by altering its initial conditions by changing TSF_0 (cell F7). Set TSF_0 to 1, and record a tick mark in the box in the column corresponding to the value of TSF_{100} (cell K24) in Table 12.1 which can be viewed using Adobe Acrobat Reader and printed from the CD. In other words, if $TSF_0 = 1$ and the model outputs a 2 as its final state, you would add a '|' to the box on the grid in row 1 and the column numbered 2. Recalculate the model four more times, recording each model outcome in the appropriate box on the same row of the table.

2. Next, increase TSF_0 in increments of 6 and recalculate the model five times at each increment, recording results in Table 12.1, using the appropriate row down the table.

Question 1.3. Do some values of TSF_{100} appear to be more common than others?

Question 1.4. What does increasing TSF_0 represent ecologically?

Question 1.5. As TSF_0 is increased, what tends to happen to TSF_{100}?

EXERCISE 1.3

1. Now, you'll set the **Initial Time Since Fire** of the model TSF_0 (cell F7) back to 1 and vary the **Probability of Fire Ignition (cell F4).** Set P_{ignit} to 0.1, and then increase its value by increments of 0.1 until

you reach 1.0. At each increment recalculate the model eight times. In Table 12.2 on the CD, mark the number of times each **Time Since Fire after 100 years** (cell K24) occurs at each P_{ignit} setting (i.e., when P_{ignit} = 0.3, and the model outputs 11 as its final state, you would add a '|' tick mark to the box on the grid in row 0.3, column 11.) This tabulation of model outcomes will give you a rough idea of how TSF_{100} varies as P_{ignit} is altered.

Question 1.6. Can you detect any changes in the distribution of "Time Since Fire after 100 years" values as P_{ignit} is increased?

Question 1.7. Is there a range of **Probability of Fire Ignitions** that exhibits a greater variety in outcomes? What values of P_{ignit} define this range (i.e., between 0.1 and 0.3)?

You have now examined the basic structure of the model. The next portion of Part 1 shows you how you can analyze the behavior of this model in a more sophisticated fashion.

Transition Matrix

Now switch to the second page of the spreadsheet, **Transition Matrix,** by clicking on the tabs at the bottom of the spreadsheet window.

This page shows the same model as the first page with the same input parameters, except that it does not require an initial condition setting. Here, you see the transition matrix that describes how the state of the model changes over time. A transition matrix is a matrix of probabilities of transitions from one state of the model to another, such as the probability of the transition from $TSF = 1$ to $TSF = 2$ (the site not burning), or from $TSF = 5$ to $TSF = 1$ (the site burning).

The matrix corresponding to the input parameters is shown below the title at cell F10. Go to the first entry in the matrix (cell G13), which calculates the probability that a site that hasn't burned for 1 year will remain in that state (that is, it will burn this year). Cell H13 contains the probability that a site that is 1 year old will become 2 years old (which is P_{nofire}, the probability that the site will not burn this year).

Similarly, the probability that a site that is 2 years old ($TSF = 2$) will become 1 year old ($TSF = 1$) is P_{fire}. This probability is shown in cell G14. Cell H14, which represents the probability of moving from $TSF = 2$ to $TSF = 2$, is zero. This probability is zero because there is no chance that a 2-year-old site will remain 2 years old; it will either get older (and become $TSF = 3$, cell I14) or burn and become 1 year old ($TSF = 1$).

Examining the overall pattern of the matrix, you should see that the first column in this matrix is the probability of sites of different ages burning. The offset diagonal in the matrix is the probability of sites becoming older when they do not burn. This transition matrix is displayed in Figure 12.5 lower on the spreadsheet (cell C39). **Figure 12.5a, Transition Matrix,** displays the entire transition matrix, while **Figure 12.5b, Expanded Area,** shows a magnification of the transitions between the sites that have not burnt for 10 years or less.

EXERCISE 1.4

1. Change the spreadsheet display to a split screen (using the small tab in the top right corner of the window, just above the end of the scroll bar arrow—or search in Excel help under "split box") so that you can see both **Figures 12.5a and b** and the input area (particularly cell F4).
2. Change P_{ignit} (cell F4) from 0.05 to 1.0, in 0.05 increments.

Question 1.8. What changes do you see occurring in the transition matrix? Explain what these changes mean in terms of model dynamics. Specifically, what is happening to the chance of a site burning; that is, what is happening to different *TSF* values shifting to *TSF* value 1?

Question 1.9. What is happening to the probability of a site not burning (i.e., advancing up the diagonal of the matrix as *TSF* increases by 1 year)?

If you know the transition matrix of an ecological system, you can conduct analyses that allow you to understand its dynamics. You will use this transition matrix to analyze the site model in the next section.

Probabilistic Site Model

Click on the tabs at the bottom on the spreadsheet window to switch to the third page of the spreadsheet, the **Probabilistic Site Model.** The Probabilistic Site Model is similar to the original Site Model with which you are familiar. While the Site Model stochastically simulates the dynamics of a specific site's *TSF* over time, the Probabilistic Site Model is more general. It calculates the probability of different *TSF* states over time. In other words, the Site Model tracks one possible site history, while the Probabilistic Site Model calculates the likelihood of all possible site histories. The probabilistic model uses the transition matrix, from sheet 2 of the spreadsheet, to simulate changes in the probability distribution of model states over time.

The output of the probabilistic model is displayed in two new graphs. The graph in **Figure 12.6, Probability Distribution of TSF States over Time** (cell I2), shows how the probability distribution of sites changes over time. **Figure 12.7, Probability Distribution of TSF States at 100yrs** (cell D32), more specifically shows the probability of the different vegetation states after 100 years of simulation. These outputs are, respectively, analogous to **Figure 12.3, Site Dynamics of TSF** and TSF_{100} in the Site Model.

The probabilistic model works by considering the condition of the site not as a specific value, but as a probability distribution of values. Each year these probabilities are multiplied by the transition matrix to produce an updated set of probabilities. In this way the changes in the probability distribution in the model can be tracked over time.

EXERCISE 1.5

1. You may be able to see both the input area and the graphs if you change the window display to a split screen. However, whether you can do this or not depends on the size of your computer's screen.

2. Change the **Probability of Fire Ignition** (cell F4) from 0.05 to 1.0 in increments of 0.05. Observe the changes in the model output in **Figures 12.6** and **12.7**. As you make these incremental changes in P_{ignit}, use the output displayed in **Figure 12.7** to answer the following questions.

Question 1.10. In what range of P_{ignit} is there a greater than 5% probability that after 100 years *TSF* will be 25 years?

Question 1.11. In what range of P_{ignit} is there a greater than 5% probability that after 100 years the *TSF* will be 1 year?

Question 1.12. In what range of P_{ignit} values is there a greater than 5% probability that *TSF* could be both 1 year and 25 years?

EXERCISE 1.6

Set P_{ignit} to 0.40. Examine **Figure 12.6, Probability Distribution of TSF States over Time.** You should be able to use the shape of these probability curves to predict what will happen to the model further into the future.

Question 1.13. Will the probability that *TSF* is 1 increase or decrease at time 200 years, compared to the probability that *TSF* is 1 at 100 years? Explain why.

Question 1.14. Similarly, will the probability of state 25 increase or decrease in 200 years? Explain why.

These analyses allow you to more fully understand the range of the model's behavior. Further, more precise analyses are possible, but they cannot be performed easily in Excel.

Calculation of Stable States

Move to the fourth and final page in the site.xls spreadsheet, **Stable States**. Unlike the other pages, Stable States does not contain a model. It contains the results of a mathematical analysis of the model. The Probabilistic Site Model examined the dynamics of a site over 100 years, but you can use matrix algebra to further analyze the transition matrix and discover the stable states that the model converges toward over infinite time. The Stable States page displays the probability distributions that the model converges toward after an infinitely long period of time over a broad range of P_{ignit} values. By varying other variables, one could obtain other stable probability distributions of the model's *TSF* values.

The stable state of the probability distribution in the model can be calculated from the dominant eigenvectors of the transformation matrix. An eigenvector of a matrix is a vector that is not distorted when it is multiplied by that matrix. This is important because each year the model multiplies the probability distribution of a state by the transition matrix, so over time, the repeated multiplication gradually shapes the vector that contains the probability distribution of states toward the dominant eigenvector of the matrix. This occurs because this vector is stable when it is multiplied by the transition matrix, while other vectors are unstable. This may be familiar to you from Chapter 4, Introduction to Markov Models.

Eigenvectors cannot be easily calculated for large matrices in Excel, but these eigenvectors and the corresponding steady state probability distribution of the model were generated using the program MatLab, and matrix analysis methods. For a detailed discussion of these methods, consult Caswell (1989) and Roughgarden (1998).

The results for this model are displayed in **Figure 12.8a, Stable TSF States according to the Probability of Fire Ignition** (cell C3) and **Figure 12.8b, Magnification of P_{ignit} values: $0.45 < P_{ignit} < 0.85$** (cell T3). These graphs display results from the matrix in cells B33 through CX58. Examine these results, while recalling how variation in the **Probability of Fire Ignition** altered the probability of different vegetation states after 100 years.

EXERCISE 1.7

Examining **Figure 12.8b**, observe that there is a greater than 10% chance of either stable state occurring when P_{ignit} ranges between 0.6 and 0.75. With the Probabilistic Site Model you obtained results after 100 years of probabilistic simulation that also indicated a range of probability values that possessed alternative stable states. However, the simulated results and these analytical results differ.

Question 1.15. Suggest why the analytical results at infinity, shown in **Figure 12.8a and b,** differ from your results in Exercise 1.6.

Question 1.16. In what situations could these differences be ecologically significant?

Resilience

The ability of a state to persist despite disruption is termed *resilience* (Holling, 1973). Ecosystems are resilient when ecological interactions within the ecosystem reinforce one another, dampening disruptions. In the example here, fire encourages the growth of easily combustible vegetation, which in turn encourages fire. **Ecological resilience** assumes that an ecosystem can exist in alternative self-maintaining or "stable" states and is a measure of the amount of change or disruption that is required to transform an ecosystem from being maintained by one set of mutually reinforcing processes and structures to a different set of processes and structures. The concept of ecological resilience can help ecologists and managers focus on transitions between states, defined by sets of organizing processes and structures, as well as the likelihood of such an occurrence (Berkes and Folke, 1998). The opposite of resilience is brittleness. A **brittle** ecosystem is vulnerable to disruption and reorganization (Gunderson et al., 1995).

EXERCISE 1.8

The ability of young and old states to persist changes as P_{ignit} is altered. At low values of P_{ignit}, only infrequently burnt areas ($TSF = 25$) are likely to persist, while when P_{ignit} is nearly 1.0, only frequently burned areas ($TSF \sim 1$) are likely to persist. Therefore, at low values of P_{ignit}, unburned sites are very resilient, while recently burned sites have little resilience and are likely to convert to unburned sites. Using Figure 12.8, qualitatively plot changes in the "resilience" of both frequently burned sites (TSF values near 1) and infrequently burned sites (TSF values near 25) on the y-axis, against changes in P_{ignit} on the x-axis.

Part 2. Landscape Resilience

Fire is a spatial process. A burning site can ignite its combustible neighbors, spreading fire across a landscape; thus, the spread of fire across a landscape is shaped by that landscape. Areas that are unlikely to burn will impede the spread of fire, while areas that are likely to burn will encourage the spread of fire (Turner et al., 1989). Assuming that the probability of fire ignition remains constant, sites will burn more frequently if fires are able to spread across large areas than they would if fire has difficulty spreading. This suggest a hypothesis that patches of longleaf pine must be above a minimum size if they are to burn frequently enough to prevent conversion to hardwoods.

Landscape Model   [*Ignore error message when opening file.*]

The spread of fire across a heterogeneous landscape is not represented in the Site Model. However, a Landscape Model can be created by connecting several Site Models. The **Landscape Model** is contained in a completely separate file (**land.xls**). This spreadsheet contains a set of 12 connected site models, allowing the investigation of spatial phenomena across a landscape.

MODEL INPUT

The landscape model consists of a line of connected site models. The input parameters include all the inputs of the Site Model, along with a few modifications. The chief difference is that rather than one initial Time Since Fire, this model contains 12 separate TSF_0 initial conditions, representing the heterogeneity that might be expected across an entire landscape.

From the Site Model: As in the other spreadsheets, the probability of a site of a specific age burning is defined by the variables P_0 (cell D4), R (cell D5),

K (cell D6), and the curve generated by these parameters is displayed in **Figure 12.9, Probability of Combustion** (cell C8).

New in Landscape Model: While the landscape model functions almost identically to the site model, the difference is the addition of one new process in the landscape model—the spread of fire from neighboring sites. If fire burns a cell, it can then ignite its neighbors. In the landscape model, the top and bottom of the column of sites are connected. Site 1 (cell K4) is a neighbor of both site 2 (cell K5) and site 12 (cell K15), and site 12 is a neighbor of site 1 (cell K4) and site 11 (cell K14). This connection makes all of the sites equivalent—each site has the same number of neighbors, eliminating edge effects from the model.

The **Probability of Fire Ignition** sets the probability that any one site is ignited (cell G4). In the Landscape Model, fire ignition functions slightly differently than it does in the Site Model. Recall in the Site Model that P_{ignit} is the probability that a site is ignited. In the Landscape Model, since there are 12 separate sites, there may be more than one ignited site in the model per year. Thus, the expected number of ignitions per year over the entire landscape (for a given P_{ignit}) is shown in cell G5; confirm that this value is P_{ignit} multiplied by the number of sites in the landscape.

At the right of the gray parameter box, the green column, *Initial TSFs*, defines the initial conditions of the 12 sites (cells K4–K15). These values correspond to the Initial Time Since Fire in the original Site Model, except that rather than one value defining the model, the model is composed of a different value for each of the 12 sites. Thus, all of the values in this column together describe the heterogeneity of initial vegetation (*TSF*) across the entire landscape.

Model Output

Directly to the right of the Initial TSFs input column, the spatial dynamics of the model are displayed graphically in **Figure 12.10, Landscape Dynamics of TSF** (cell M2). The model is arranged so that the sites are oriented vertically, and the calculations of the model's yearly dynamics are also done vertically. Each column represents one year's dynamics in the model. You should set the initial value of all sites (cells K4–K15) to 1, and then change site 5 (cell K8) from 1 to 25. Note the change in the graph. Change the same site to 11. Recalculate the model several times (using the F9 key) and observe the variation in the model output.

The graph on the middle left-hand side of spreadsheet, **Figure 12.11, Distribution of TSF States after 100 Years** (cell C18), shows the distribution of the sites' vegetation state after 100 years of simulation. **Figure 12.12, Fire Ignition and Area Burned** (cell M18), shows the corresponding graph of the proportion of the landscape in which fires were ignited, and the proportion of the area burned every year. **Figure 12.13, Dynamics of Individual Sites** (cell M27), displays the changes in each of the individual sites over 100 years.

Now that you have seen how the model input parameters are entered and the model results are displayed, you should examine how the model actually functions.

How Does the Model Work?

The calculation of the model is done at the right of the spreadsheet, beyond the graphs. The calculations begin at column AE, with each column representing the dynamics of 1 year.

Initial Conditions. Cells AF5 to AF16 contain the initial conditions of the model. Click on cell AF5 and you will observe that it links to cell K4, the cell in which the initial conditions for the first site are set. If you click on AF6, you will observe that it links to cell K5, the initial conditions for the second site, and so on. From these initial conditions the behavior of the model during the year is calculated.

Probability of Combustion. Moving down the column to the next block of cells brings you to the calculation of each site's probability of combustion. Cells AF19 to AF30 contain the P_{comb} of sites 1 to 12. Click on site AF19 and observe that these probabilities are calculated using the fire spread function that is defined by the parameters P_0 (cell D4), R (cell D5), K (cell D6), and the initial *TSF* in each of the sites (defined in cells AF5 to AF16). This function is identical to that used in the original Site Model. Click on AF20. Notice that the function is the same. The same variables are used with the exception that site 1 (AF5) is replaced by site 2 (AF6).

Probability of Ignition. The next block of cells down, cells AF32 to AF43, is a stochastic portion of the model. These cells calculate which sites are ignited by generating a pseudorandom number between 0 and 1 for each site. If the number is less than the probability defined by the **Probability of Fire Ignition** parameter (cell G4), then the site is combustible. If the number is higher than P_{ignit}, then the site is not ignited that year. The pseudorandom numbers are uniformly distributed. Thus, when the probability of being ignited is 0.45, then on average, 45 times in 100, the site will be ignited. Ignited sites are designed with 1's, while nonignited sites are marked with 0's.

Now recalculate the model. Note that the probabilities in cells AF19–AF30 remain constant because they are calculated from the models initial values, which are constant. However, notice that whether the sites are ignited or not (cells AF32–AF43), changes as the model is recalculated because the ignition of the sites is a random variable.

Burn Mask Cells. Further below, the burn mask cells (cells AF47–AF58) calculate whether a site is combustible. This is done by determining if a generated pseudorandom number is lower than the probability required for the site to burn (as defined in cells AF19–AF30). If the number is less than the probability value, then the site can burn. If it is higher than the probability value, then it will not burn. For example, if the probability of a site burning is 0.97, then it will on average burn 97 times out of 100. Combustible sites are designed with 1's, while noncombustible sites are marked with 0's. Click on cell AF47 and note the comparison of the random number to the site 1 probability of burn (AF19). Click on AF48 and note that it is calculated identically, but uses the values from site 2 (AF20). Now recalculate the model and note how these values change as variation in the pseudorandom numbers results in sites being combustible or not.

Fire Initiation. The initiation cells AF60–AF71 use the ignition cells and the burn mask cells to determine what sites are successfully ignited, which sites are combustible, and which sites are not combustible. The initiation cells calculate if a cell is burning—that is, if it is both combustible and ignited. If a site is burning, then its value is set to 2. Other site values are copied from the burn mask, where 1's are combustible and 0's are not. Click on a couple of the AF60–AF71 cells and observe the formulas, then recalculate the model, again noting how the values change as the model calculates different pseudorandom numbers.

Spread. The next sets of cells, spread 1, starts to calculate the spread to adjacent sites. These cells convert sites that are burning (2) into sites that are extinguished (−1). If a burning site is next to a combustible site (1), the combustible site is ignited (2). This process is done first in cells AF73–AF84, and then repeated 12 times in the different sets of cells, in spread 2 through spread 12. Spread 12 is the maximum distance fire can spread.

Most of the time it isn't necessary to calculate spread this many times, but 12 calculations are necessary to calculate the extreme possible cases, such as when fire spreads in one direction across the entire landscape. If you recalculate the model, you will observe that there are seldom burning fires (2's) in the lower spread calculation regions—below spread 6 (cell AF149). However, with specific configurations of the model, such a spread is possible, and therefore these calculations are necessary.

Burn Area. After a fire has been allowed to spread and then extinguish itself, the consequences of the fire are calculated. The total number of sites that burned during the spread of fire are calculated in the Burn Area range (cells AF229–AF240). All cells that are −1 in the spread 12 section are marked as 1 (burned) here, and all other cells are marked 0. Click on cell AF229 and you can see the simple logic statement that does this calculation.

Update. The new values for the sites are calculated using the data in the Burn Area cells. Update cells (AF242–AF253) are set to 1 if they were burned (that is, their site was a 1 in the Burn Area cells). In other sites the initial value (AF5–AF16) of the site is increased by 1, unless the site is already at 25 years since fire (in which case it remains at 25 years). Examine the logic statements used to do this in cells AF242–AF253.

Statistics. Fire statistics are calculated below the update cells. The fire count section collects the data to create Figure 12.12. These are summations of the number of sites that burned, and the number of sites in which fires were initiated. You can click on these cells, AF255–AF257, to examine these formulas. The update cells are used to set the initial values in the next column, AG. Click on AG5–AG16 to observe how they read these update values, from the bottom of the previous column.

The entire fire ignition, spread, and vegetation change process is repeated in each column. The results of the calculations in a column are then used to start the calculations for the next year, in the next column. Now that you understand the calculation of the model, you can use it to conduct a few exercises.

EXERCISE 2.1

The ability of fire to spread across sites in the landscape model alters the model's dynamics as compared to the dynamics of only a single site.

1. To demonstrate this, you should first confirm that the model's input parameters are set to their standard values. The fire spread function parameters should be P_0 (cell D4): 0.00125, R (cell D5): 0.65, K (cell D6): 0.975, and **Probability of Fire Ignition** (cell G4): 0.45, and Time Since Fire (cells K4–K15): 1.

2. If you recalculate the model several times, you will notice that unlike the Site Model, a P_{ignit} of 0.45 results in the Landscape Model always burning and remaining at $TSF = 1$.

3. Compare the Landscape Model's P_{ignit} that begins to produce sites that have TSF of 25 years, with P_{ignit} from **Figure 12.8, Stable TSF States according to the Probability of Fire Ignition,** in the Stable States sheet of site.xls.

Question 2.1. Lower the P_{ignit} until you just begin to get model runs where all of the landscape enters the old state, with a TSF of 25 years or greater. At what P_{ignit} does this begin to happen?

Question 2.2. Why is there is a difference between the P_{ignit} values required to produce sites having a $TSF = 25$ on the Landscape Model and those in the Site Model from the Stable States page?

EXERCISE 2.2

The landscape model demonstrates the importance of the spatial distribution of vegetation in determining landscape dynamics. For example, the same initial mix of sites can produce different results depending on the spatial distribution of sites. To demonstrate this, you will use the model to conduct a four-part simulation experiment.

1. Set P_{ignit} to 0.15. Set the values in site 4 and site 5 (cells K7 and K8) to 25, and all other sites to 1. Run the model five times. Record the cumulative distribution of states at 100 years (from **Figure 12.11, Dis-**

tribution of TSF States after 100 Years) in **Table 12.3** from the CD for each run. In other words, for each run, count and record the number of sites in each *TSF* class (e.g., four sites that are *TSF* 1–2). Also, observe the model dynamics in **Figure 12.10, Landscape Dynamics of TSF.**

2. Move the 25-year *TSF* sites one site apart. Set the value in site 5 (cell K8) back to 1 and set the value of site 6 (cell K9) to 25, while keeping site 4 at 25, and all other sites at 1. Again, run the model five times, recording the cumulative distribution of states at 100 years (shown in **Figure 12.11**) in **Table 12.4** from the CD. Notice the dynamics of sites 5 and 6 versus sites 1–3 and 8–12 revealed in **Figure 12.10.**

3. Move the 25-year *TSF* sites three sites apart. Set the value in site 6 (cell K9) back to 1 and set the value of site 8 (cell K11) to 25, keeping all other sites the same. Once more, run the model five times, recording the cumulative distribution of states at 100 years, shown in **Figure 12.11**, in **Table 12.5** from the CD. Notice the difference in the behavior of the model as shown in **Figure 12.10** with these settings and the other model runs.

4. Finally, move the 25-year *TSF* sites five sites apart. Set the value in site 8 (cell K11) back to 1 and set the value of site 10 (cell K13) to 25, while keeping site 4 at 25, and the other sites at 1. Once more, run the model five times, recording the cumulative distribution of states at 100 years, shown in **Figure 12.11**, in **Table 12.6** on the CD. Note the difference in the behavior of the model as shown in **Figure 12.10** with these settings and the other model runs.

Question 2.3. The model runs that you conducted in this exercise all begin with the same proportion of old and young sites; however, they have different mixes of states after 100 years. What effect does the location of old sites have on the spread of fire?

Question 2.4. What effect does the location of old sites have on the dynamics and the state of the model after 100 years?

Question 2.5. Based on your experiments with different landscape patterns (old sites separated by 0, 1, 3, and 5 sites), what does the model suggest to you about the resilience of different landscape patterns? What patterns of old and young sites are more resilient (i.e., more likely to persist)?

Question 2.6. What patterns of old and young sites are less resilient (i.e., brittle or unlikely to persist)?

Question 2.7. What does this model suggest about the impact of landscape fragmentation on ecological processes such as fire?

CONCLUSIONS

The models investigated in this lab demonstrate how an ecological system can exist in alternative stable states and how the dynamics of alternative stable states change when moving from a site to an entire landscape. In this example, a site could persist as either a young, frequently burned site (dominated by longleaf pine) or an old, infrequently burned site (dominated by oaks). Interestingly, however, the dynamics of a particular site changed when the site was placed within the context of its landscape, and the dynamics of a site greatly depended on the *specific* landscape context in which it was placed. Also, the dynamics of any single site were quite different from those of the entire landscape. When sites were grouped together to create a landscape, a site tended to push its neighboring sites toward its own particular state, reducing the range of system parameters for which alternative stable states could persist.

In the forests of northern Florida, the ability of fire to spread, and consequently the rate at which patches of hardwood or pine either grow or shrink, is influenced by the pattern of hardwoods and pine across the landscape. The combination of spatial heterogeneity and positive feedbacks makes the dynamics of forests difficult to predict from the study of a local site, since the processes that control a site can be affected by the properties of its neighbors. However, the use of models allows one to examine how the configuration of habitats across a landscape can affect the resilience of landscapes in the face of disturbance. Furthermore, while the models used in this lab are based on the forest dynamics of northern Florida, many ecosystems exhibit alternative stable states (Dublin et al., 1990; Knowlton, 1992; Scheffer et al., 1993). The dynamics of these ecosystems can be analyzed using similar models and techniques.

BIBLIOGRAPHY

Note. An asterisk preceding the entry indicates that it is a suggested reading.

BERKES, F., AND C. FOLKE, EDS. 1998. *Linking Ecological and Social Systems.* Cambridge University Press, Cambridge, UK.

*CASWELL, H. 1989. *Matrix Population Models: Construction, Analysis, and Interpretation.* Sinauer Associates, Sunderland, Masscahusetts. A thorough guide to the development, use, and analysis of matrix models.

*DUBLIN, H. T., A. R. E. SINCLAIR, AND J. MCGLADE. 1990. Elephants and fire as causes of multiple stable states in the Serengeti-Mara woodlands. *Journal of Animal Ecology* 59:1147–1164. Describes how fire and elephants regulate the multiple stable vegetation states in the African savanna. This system is similar to that found in northern Florida, except that for over 11,000 years Florida has not had any elephants or other mega-herbivores.

*GLITZENSTEIN, J. S., W. J. PLATT, AND D. R. STRENG. 1995. Effects of fire regime and habitat on tree dynamics in north Florida longleaf pine savannas. *Ecological Mono-*

graphs 65:441–476. Experimental investigation of the complex effects of fire on tree dynamics.

GUNDERSON, L. H., C. S. HOLLING, AND S. S. LIGHT, EDS. 1995. *Barriers and Bridges to the Renewal of Ecosystems and Institutions.* Columbia University Press, New York.

*HOLLING, C. S. 1973. Resilience and stability of ecological systems. *Annual Review of Ecology and Systematics* 4:1–23. Introduces and explains the concept of ecological resilience.

*KNOWLTON, N. 1992. Thresholds and multiple stable states in coral reef communities. *American Zoologist* 32:674–682. Reviews the literature on multiple stable states in coral reef communities.

PETERSON, G.D. 1999. *Contagious Disturbance and Ecological Resilience.* University of Florida—Gainesville. Dissertation. Gainesville, Florida.

*Rebertus, A. J., G. B. Williamson, and E. B. Moser. 1989. Longleaf pine pyrogenicity and turkey oak mortality in Florida xeric sandhills. *Ecology* 70:60–70. Interaction of fire, oak, and longleaf pine.

*ROUGHGARDEN, J. 1998. *Primer of Ecological Theory.* Prentice Hall, Upper Saddle River, New Jersey. Includes a quick introduction to the use and analysis of matrix models.

*SCHEFFER, M., S. H. HOSPER, M.-L. MEIJER, B. MOSS, AND E. JEPPESEN. 1993. Alternative equilibria in shallow lakes. *Trends in Ecology and Evolution* 8:275–279. A review paper on alternative states in shallow lakes.

*TURNER, M. G., R. H. GARDNER, V. H. DALE, AND R. V. O'NEILL. 1989. Predicting the spread of disturbance across heterogeneous landscapes. *Oikos* 55:121–129. Provides a general model for the spread of disturbance across a patterned landscape.

ORGANISM RESPONSE TO LANDSCAPE PATTERN

The response of organisms to landscape pattern is one of the better-developed themes in landscape ecology, and this section reflects the diversity of approaches to this topic. Chapter 13, Interpreting Landscape Patterns from Organism-Based Perspectives, is designed to show how different organisms may perceive the same landscape in very different ways, and examines the implications for mapping species' habitat requirements. Chapter 14, Landscape Context, examines the influence of landscape features *surrounding* a patch on the species present within a patch. This concept of landscape context is an important contribution of landscape ecology. Landscape ecologists have also developed several lines of theoretical inquiry to understand the effects of landscape structure on animal movement and population persistence. Chapter 15, Landscape Connectivity and Metapopulation Dynamics, links neutral landscape models and metapopulation theory to compare population persistence of different organisms in fragmented and clumpy landscapes. Last, the development of individual-based models has been important for addressing issues of organism movement and population persistence in an applied setting. Chapter 16, Individual-Based Habitat Modeling: The Bachman's Sparrow, enables students to use BACHMAP, a spatially explicit, individual-based model, to evaluate the persistence of the endangered Bachman's sparrow under different landscape management scenarios.

CHAPTER 13

INTERPRETING LANDSCAPE PATTERNS FROM ORGANISM-BASED PERSPECTIVES

SCOTT M. PEARSON

OBJECTIVES

Conservation is a geographic problem because one of the greatest threats to biodiversity is habitat loss and fragmentation. Management plans, with purposes ranging from protecting endangered species to planning sustainable timber harvests, use maps to represent current landscape patterns and to evaluate management options. Obviously, the accuracy of these maps can affect the success of the decision-making process and the implementation of management plans. For ecosystem-level or species-based management, maps must reflect patterns relevant to ecosystem processes and species of interest. However, maps are human products; therefore, they have an implicit anthropocentric perspective. Most tend to record information pertinent to human needs and economic systems (e.g., land cover, roads, towns, political boundaries). For example, national forests have "stand maps" that record the age and type of forest stands (e.g., oak-hickory, cove hardwood, oak-pine). However, this classification system is based on merchantable timber and may not reflect stand characteristics important to wildlife or nontimber vegetation. For effective conservation, maps are needed that reflect the needs of ecosystems and nonhuman species. The objectives of this exercise are to

1. illustrate how landscape patterns, recorded on land-cover maps, can be interpreted from the perspective of different species; and

2. help students understand the uses and limitations of spatial data for producing habitat maps.

In Exercise 1, you will examine the potential limitations of habitat mapping due to the resolution of the spatial data used. In Exercise 2, you will interpret landscape patterns from a non-anthropocentric perspective and construct habitat maps for five species. In Exercise 3, land-cover and topographic data will be combined to represent the habitat of two additional species. Each team of students will need several different colors of pencils, markers or highlighter pens; a calculator; and five copies of Figure 13.1 for Exercise 2. A transparency of Figure 13.1 and a paper copy of Figure 13.2 are also needed for Exercise 3 as this part involves overlaying Figures 13.1 and 13.2. Extra copies (for printing) of the figures are included on the CD in the directory for this lab and can be viewed using Adobe Acrobat Reader software which is available for free from the World Wide Web.

INTRODUCTION

One of the challenges of ecosystem management is understanding the effects of landscape-level changes on biological diversity. Both worldwide and in the United States, land cover is altered principally by direct human use through agriculture, pasture, forestry, and development. Land-use patterns affect both terrestrial and aquatic systems (Reiners et al., 1994; Cooper, 1995) and influence biodiversity for several reasons (Turner et al., 1998). First, land-use activities may alter the relative abundances of natural habitats and result in the establishment of new land-cover types. Species richness may be enhanced by the addition of new cover types, but natural habitats are often reduced, leaving less area available for native species (Walker, 1992). Exotic species may also become established and outcompete the native biota. Second, the spatial pattern of habitats may be altered, often resulting in fragmentation of once-continuous habitat. Projecting patterns of species presence and abundance in changing landscapes remains a key challenge in sustaining biodiversity (Lubchenco, 1995; Hansen et al., 1995), and clearly, the conservation of native species and their habitat involves a landscape-level approach (Franklin, 1993, 1994; Tracy and Brussard, 1994).

Ecologists use a variety of terms to describe the spatial pattern of habitat, such as *patch, connected, fragmented, edge,* and *edge effects.* For the purposes of this exercise, **habitat** refers to sites having appropriate levels of the biotic (e.g., prey items, mutualists) and abiotic (e.g., moisture, temperature, light, nutrients) features required by a given species. A **patch** is a contiguous region of the same habitat type. In raster land-cover maps, such as Figure 13.1, patches are usually identified as clusters of contiguous map cells of the same cover type. When the habitat exists in the landscape as a few large patches, the habitat is said to be highly **connected.** Habitat is considered to be **fragmented** when the habitat occurs in a large number of small patches. Land-cover change that results in habitat loss often causes habitat fragmentation by breaking up the few large patches into an increased number of small ones. The boundary between two habitat types is called an **edge**. Within a patch, factors such as temperature, moisture, food resources, and the abundance of predators may be different near the edges of a patch as compared to the **interior** of a patch (Matlack, 1993). Changes in ecological factors near patch boundaries are termed **edge effects**. Because smaller habitat patches have a greater edge-to-interior ratio, edge effects tend to be more prevalent in small patches. Thus, habitat fragmentation, by creating small patches, can increase the amount of edges and edge effects throughout a landscape.

Depending on their habitat requirements and life-history attributes, species may respond quite differently to habitat loss and fragmentation. Changes that favor one species may reduce habitat for others. The abundance and spatial pattern of habitat in a landscape may vary among species that have different habitat requirements (e.g., preferences for late versus early successional stages). Moreover, life-history attributes, such as area requirements and dispersal ability, can interact with the spatial pattern of habitat (i.e., fragmented vs. connected) to affect population dynamics on a landscape. Therefore, an organism-based perspective (e.g., Wiens, 1989; Pearson et al., 1996) is needed to estimate the effects of landscape pattern on nonhuman species. This perspective should also acknowledge that species may perceive the habitat in different ways. An eagle "sees" the landscape, its habitat patches, and resources in a very different way than a beetle does. Ecologists should strive to view the landscape from the perspective of the species of interest. The following exercises address the importance of grain size on habitat mapping, show how the same map can be interpreted differently for different species, and integrate topography with land cover in mapping as ways to illustrate the importance of interpreting landscape patterns from an organismal perspective.

Study Area

The study area is located north of Asheville, North Carolina, in the Southern Blue Ridge Province of the Southern Appalachian Mountains, with elevations ranging from 660 to 1100 meters above sea level. This mountainous landscape is dominated by forests, particularly in the steeper areas at higher elevation, while many of the valleys have been cleared for agricultural and residential land uses. The study area also includes urban areas of a small town, Mars Hill, having a human population of 2200 (Figure 13.1). The mountainous areas north of the town are heavily forested. The land-cover types we will consider here are forest, nonforest, and urban. The forests of this area are mostly temperate deciduous forests interspersed with occasional conifers (e.g., pines and hemlocks). The nonforest land cover includes pastures, row crops, and lawns. Urban land cover includes cells with a high concentration of buildings and pavement. Landscape metrics for these land covers are listed in Table 13.1.

FIGURE 13.1
Land-cover map of study
area. Thin lines are 10-m
contours; heavier lines represent streams. Each cell is
90 × 90 m in area. The
spatial extent of the map
is 19.7 km^2 (3.78 ×
5.22 km). A full size version of this map can be
printed from the CD.

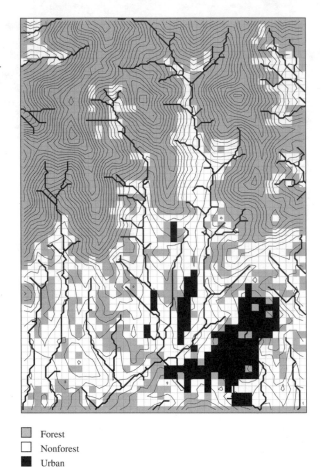

◼ Forest
☐ Nonforest
■ Urban

TABLE 13.1
LANDSCAPE PATTERN METRICS FOR FIGURE 13.1. AREA AND PATCH SIZE
ARE MEASURED IN THE NUMBER OF 90 × 90-M CELLS.

Land Cover	Total Area (cells)	Number of Patches	Mean Patch Size (cells)	Size of Largest Patch (cells)
Forest	1349	37	23.8	1133
Nonforest	910	14	42.6	791
Urban	135	6	14.7	25

EXERCISES

EXERCISE 1
The Importance of Grain Size

Habitat maps are produced from spatial data that have a certain spatial resolution. For example, the cell size of a **Thematic Mapper (TM)** image is typically 30 × 30 meters. The land-cover map used for this exercise was produced from a 1991 TM image (Figure 13.1) and is stored in raster (i.e., grid-based) format. TM is a type of satellite used to gather digital photographs. Regions within the photograph are classified into discrete land-cover categories. The resulting land-cover classes can be represented in **raster** format, using a large rectangular array of numbers, similar to a checkerboard. Each number represents the land-cover type for the cell. One pixel (cell) of a TM image usually represents a 30 × 30-meter area on the ground, so the "size" of a cell in the derived map cannot be smaller than this size; however, during land-cover classification, cells may be aggregated into a more coarse resolution (e.g., 90 × 90 m). The size of an individual map cell determines the spatial resolution or **grain** of the map; the spatial **extent** of the map is its total area (e.g., N-S distance × W-E distance).

The value of a cell in a land-cover map may represent either (1) a point measured in the middle of the cell or (2) an average of the conditions in cells—thus, cells in a raster habitat map may contain much microsite variability that is not captured by the cell value. **Microsites** are the small or fine-scale locations within a map cell. For a mayfly species capable of living in small creeks, a microsite for this insect would be in the creek (as opposed to on top of a dry rock). Because a map cell may cover such a large area (relative to an insect), many microsites within the cell may not be suitable. Furthermore, because a small creek (2–3 m across) represents such a small proportion of the land area, it would not show up in a TM image; however, larger streams, rivers, and ponds would be recorded.

Question 1. Consider an insect species that requires small headwater streams. Can habitat for this species be accurately mapped using data with a 90 × 90-meter resolution? Justify your answer. If you answered yes, then explain your assumption(s) about the relationship between the cell values and the abundance of suitable microsites.

Question 2. How does the grain of spatial data affect our ability to map species' habitats?

EXERCISE 2
Habitat Maps for Different Species

One of the challenges of ecosystem management is determining how a set of species with diverse habitat needs will be affected by landscape-level changes. Species may respond to landscape patterns in different ways depending on their habitat needs. Producing more habitat for one species by manipulating land cover may reduce habitat for others. Next, you will make habitat maps for five different species. The land-cover map (Figure 13.1) will be used to produce habitat maps using a mapping procedure based on the habitat requirements of each species (Table 13.2). Then, you will compare the abundance and spatial distribution of suitable habitats for these five species.

Impatiens capensis (jewelweed) is a native plant that thrives in spring seeps, bogs, and streamsides. This species is tolerant of shade but does best in the presence of sunlight and plenty of moisture. It is an annual that produces a large number of seeds. The seeds are borne in pods that fly apart when ripe (hence the other name of "touch-me-not").

Seiurus aurocapillus (Ovenbird) is a Neotropical migrant songbird that nests throughout the eastern United States. It nests and forages on the ground in deciduous forests and is sensitive to forest fragmentation. It is often absent from small patches of forest (< 25 ha) and occurs in greatest abundance in landscapes with a large percentage of forest. Thus, this species seems to decline with increasing forest fragmentation.

TABLE 13.2
HABITAT REQUIREMENTS AND MAPPING PROCEDURE FOR FIVE SPECIES.
HABITAT REQUIREMENTS WERE TAKEN FROM WEBSTER ET AL. (1985),
EHRLICH ET AL. (1988), WOFFORD (1989), HAMEL (1992), AND
ROBINSON ET AL. (1995).

Species	Habitat Required	Mapping Procedure
Jewelweed (*Impatiens capensis*)	Moist to hydric sites; streamsides	Cells of any land cover adjacent to or crossed by streams
Ovenbird (*Seiurus aurocapillus*)	Forest interior sites	Deciduous forest cells at least 180 m (i.e., two cells) away from nonforest and urban cells
Mountain dusky salamander (*Desmognathus ochrophaeus*)	Forests with streams	Forest cells crossed by or adjacent to streams
Indigo Bunting (*Passerina cyanea*)	Forest edge	Forest cells adjacent to nonforest and nonforest cells adjacent to forest
House mouse (*Mus musculus*)	Urban areas and nonforest	All urban cells, nonforest cells, and cells adjacent to urban cells

Desmognathus ochrophaeus (mountain dusky salamander) is one of the few members of the genus *Desmognathus* that is terrestrial. It spends a great deal of time underground, emerging to forage on the surface during warm, damp evenings. The salamander occurs along streamsides in shady, moist forests of the Southern Appalachians at low elevations. Precipitation increases with elevation in this region; thus, at elevations above 1220 meters, this species can leave the streamsides and be found in mesic forest sites.

Passerina cyanea (Indigo Bunting) is a finchlike neotropical migrant bird. This species nests in shrubs in old fields and along forest edges. Closely related to the Northern Cardinal, it feeds on insects, fruits, and seeds. The colorful males of this species are a favorite of bird watchers; the birds sing all day during late spring and summer.

Mus musculus (house mouse) is an exotic mammal introduced from Europe. It is a commensal with humans, achieving its greatest abundance in human dwellings. This species is also capable of living in the farmlands surrounding human settlements, but does not persist in forests. Not surprisingly, its abundance in the landscape is correlated with the amount and intensity of nonforest land uses. It can become a pest by destroying food items in homes and farms.

1. Your first step is to create habitat maps. Classify a land-cover map (Figure 13.1) using the mapping procedure for the species listed in Table 13.2. You can print multiple copies of the land-cover map from the CD. Use one copy of the land-cover map for each species. Color in suitable cells using a green, yellow, or red pencil, marker, or highlighter. Your instructor may create teams of students and assign a single species to each team member.

2. Next, quantify habitat abundance and pattern. A **patch** of habitat is defined as a group of *contiguous* cells of the suitable habitat. For this exercise, suitable cells contiguous on the diagonal (adjacent corners) are considered to be of the same patch, as well as those cells that share a flat edge (the eight-neighbor rule). For each patch of suitable habitat, record its size by counting the number of cells. Record the number and sizes of patches in Table 13.3, which can be printed from the CD.

3. Calculate the total area of suitable habitat (in cells) and mean patch size, and note the size of the largest patch for each species.

Using your maps, answer the following questions

Question 3. Visually compare the habitat maps for the five species. For which species does the habitat *appear* to be most connected? For which species is the habitat most fragmented?

Question 4. Which of the patch-based statistics in Table 13.3 would you use to *quantify* habitat fragmentation or connectivity?

EXERCISE 3
Using Topography to Refine Habitat Types

A quick inspection of the land-cover map (Figure 13.1) reveals patches of forest, field, and urban cover types. Within any of these covers, there is variation in environmental parameters (e.g., temperature, light, moisture) that affect habitat suitability. In this part, you will use topography to refine the habitat classification of cells with the same cover type. Realize, of course, that many other sources of spatial information may be used to refine cover types, such as geology maps, hydrology maps, and distances to certain features such as roads and buildings calculated with a Geographic Information System (GIS).

Topographic features such as terrain shape and aspect can exert a strong influence on habitat suitability in mountainous regions (Whittaker, 1956; McNab, 1996). Sites with a concave terrain shape, such as ravines, tend to accumulate water and be moist, while convex sites, such as ridges, are drier. Aspect can also affect moisture and temperature. In the northern hemisphere, north-facing aspects receive less insolation and are therefore cooler and moister than south-facing aspects. Maps of terrain shape and aspect can be derived from elevation maps.

Figure 13.2 shows a map of terrain shape for the same area as the land-cover map (Figure 13.1). Three classes of terrain shape are identified: (1) coves or ravines, (2) side slopes or flats, and (3) ridges or peaks. Cell size for Figure 13.2 is 30 × 30 meters. This map of terrain shape can be used in conjunction with the land-cover maps to identify the habitat of two trees of the deciduous forest: **basswood** (*Tilia americana*), a mesophytic species, and **scarlet oak** (*Quercus coccinea*), a xerophytic species. The cove sites are suitable for basswood, and the ridge sites are suitable for scarlet oak.

1. Map the coves and ridge areas as follows. Print out a paper copy of the map of terrain shape (Figure 13.2) from the CD. Then, create a transparency of the land-cover map (Figure 13.1). Using Figure 13.1 as an overlay on Figure 13.2, color the cells of deciduous forest that overlap at all with any cove and ridge zones. Use contrasting colors for the cove and ridge cells. *NOTE:* Print both maps on the same printer for consistent sizing results. Be sure to develop a consistent rule for classifying cells with mixed terrain (e.g., use most abundant terrain class).

2. Count the cells and patches (as in Exercise 2) and record your results in Table 13.4, which can be printed from the CD.

3. Tally the number of patches. Calculate the mean patch size and the area of the largest patch. Be sure to state your units (cells or hectares) on Table 13.4.

Use the data in Table 13.4 to answer these questions:

Question 5. Compare the relative abundance of habitat for basswood and scarlet oak. What proportion of the forest land cover is suitable for each species?

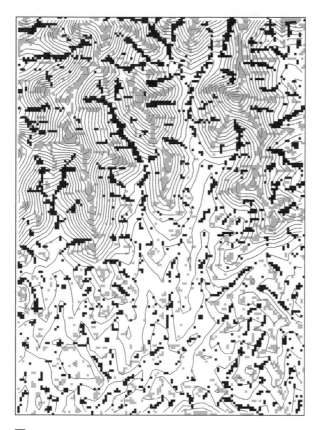

FIGURE 13.2
Map of terrain shape for study area. Thin lines are 10-m contours. The region covered and spatial extent of this map are identical to Figure 13.1; however, cells are 30 × 30 m in area. A full size version of this map can be printed from the CD.

■ Coves, ravines
□ Slopes, flats
▨ Ridges, peaks

Question 6. Compare the level of habitat fragmentation for these species. Is there a difference in the relative connectivity of cove and ridge sites in this landscape?

DISCUSSION QUESTIONS

Answer the following questions using your maps and data.

1. The Ovenbird is restricted to forest-interior cells. Compare the total number of cells of Ovenbird habitat to the total number of forest cells (Table 13.1). What percentage of the forest cells is unsuitable for the ovenbird because of edge effects?
2. Limitations in dispersal ability may prevent some species from recolonizing patches that have experienced local extinctions. Lungless salamanders, including the mountain dusky salamander, are such species because they can seldom cross dry, open land-cover types. If we assume that these salamanders cannot cross more than two

cells of *unsuitable* habitat (count cells on the diagonal as equal to cells in cardinal directions), how many of the existing patches of salamander habitat are completely isolated (with respect to potential colonists) from any other patches?
3. Refer to the habitat maps for Exercise 2.
   (a) If urban expansion in this landscape increases the extent of nonforest and urban land covers, how will each of the species discussed in Exercise 2 be affected?
   (b) Will these effects depend on the location and amount of urban expansion?
   (c) Using the lay of the land as depicted in Figure 13.2, can you predict where urban expansion is likely to occur? How?
4. Suppose that we evaluate the landscape from the perspective of another species, such as a Broad-winged Hawk (*Buteo platypterus*), which requires the same habitat as the Ovenbird but has a minimum area requirement (e.g., territory size) of 50 cells (40.5 ha).
   (a) What proportion of the patches would be too small?
   (b) What proportion of the forested cells would therefore be unsuitable?
   (c) What effect would your predicted expansion of nonforest and urban land covers have on this species?
5. Given a scenario of future urban growth *and* the potential to regulate the location of that growth, what portions of the landscape would you protect? Which species would influence your strategy?
6. What is the value of using additional spatial data, such as terrain shape, soils, geology, to refine land-cover types? What are the limitations of this approach? If the grain of the two data sets differs, what is the resolution of the resulting habitat map?

BIBLIOGRAPHY

Note. An asterisk preceding the entry indicates that it is a suggested reading.

*ANDRES, BRAD A. 1994. Coastal zone use by postbreeding shorebirds in northern Alaska. *Journal of Wildlife Management* 58:206–212. This study addresses the impact of oil development on shorebird habitats and discusses management implications. It uses a vegetation map derived from Landsat imagery.
COOPER, S. R. 1995. Chesapeake Bay watershed historical land use: Impact on water quality and diatom communities. *Ecological Applications* 5:703–723.
EHRLICH, P. R., D. S. DOBKIN, AND D. WHEYE. 1988. *The Birders Handbook: A Field Guide to the Natural History of North American Birds.* Simon & Schuster, New York.
FRANKLIN, J. F. 1993. Preserving biodiversity: Species, ecosystem, or landscapes. *Ecological Applications* 3:202–205.
FRANKLIN, J. F. 1994. Response to Tracy and Brussard. *Ecological Applications* 4:208–209.
HAMEL, P. B. 1992. *The Land Manager's Guide to Birds of the South.* The Nature Conservancy, Chapel Hill, North Carolina.
*HANSEN, A. J., S. L. GARMAN, B. MARKS AND D. L. URBAN. 1993. An approach for managing vertebrate diversity across multiple-use landscapes. *Ecological Applications* 3: 481–496. Describes an approach to balancing commodity production with the needs of terrestrial vertebrates. The approach is demonstrated for a watershed in western Oregon, balancing timber production with habitats used by birds.

HANSEN, A. J., S. L. GARMAN, J. F., WEIGAND, G. L URBAN, W. C. MCCOMB, AND M. G. RAPHAEL. 1995. Alternative silvicultural regimes in the Pacific Northwest: Simulations of ecological and economic effects. *Ecological Applications* 5:535–554.

*HE, H. S., D. J. MLADENOFF, V. C. RADELOFF, AND T. R. CROW. 1998. Integration of GIS data and classified satellite imagery for regional forest assessment. *Ecological Applications* 8:1072–1083. Demonstrates how satellite imagery and other spatial data can be combined to assess forest composition over a large region. The resulting maps could be used as input for modeling forest dynamics and to guide future management decisions.

LUBCHENCO, J. 1995. The role of science in formulating a biodiversity strategy. *BioScience*, Special supplement, pp. 7–9.

MCNAB, W. H. 1996. Classification of local- and landscape-scale ecological types in the southern Appalachian Mountains. *Environmental Monitoring and Assessment* 39:215–229.

MATLACK, G. R. 1993. Microenvironment variation within and among forest edge sites in the eastern United States. *Biological Conservation* 66:185–194.

PEARSON, S. M., M. G. TURNER, R. H. GARDNER, AND R. V. O'NEILL. 1996. An organism-based perspective of habitat fragmentation. In *Biodiversity in Managed Landscapes: Theory and Practice*, R. C. Szaro, ed. Oxford University Press, Oxford, UK.

REINERS, W. A., A. F. BOUWMAN, W. F. J. PARSONS, AND M. KELLER. 1994. Tropical rain forest conversion to pasture: Changes in vegetation and soil properties. *Ecological Applications* 4:363–377.

ROBINSON, S. K., F. R. THOMPSON III, T. M. DONOVAN, D. R. WHITEHEAD, AND J. FAABORG. 1995. Regional forest fragmentation and the nesting success of migratory birds. *Science* 267:1987–1990.

*RUDIS, V. A., AND J. B. TANSEY. 1995. Regional assessment of remote forests and black bear habitat from forest resource surveys. *Journal of Wildlife Management* 59:170–181. The authors develop a model from forest inventory data and bear censuses to develop a map of suitable habitat for 12 states in the southern United States and evaluate the relative connection and isolation of habitat patches.

*STOLZENBURG, W. 1998. When nature draws the map. *Nature Conservancy* 48:12–23. A layman's introduction to using multiple geographic data sets for mapping ecosystems. It illustrates examples from the eastern and western United States as wells as the Caribbean. The article discusses strategies for conservation efforts on a regional scale.

TRACY, C. R., AND P. R. BRUSSARD. 1994. Preserving biodiversity: Species in landscapes. *Ecological Applications* 4:205–207.

TURNER, M. G., S. R. CARPENTER, E. J. GUSTAFSON, R. J. NAIMAN, AND S. M. PEARSON. 1998. Land use. In *Status and Trends of the Nation's Biological Resources*, M. J. Mac, P. A. Opler, C. E. Puckett Haecker, and P. D. Doran, eds. U. S. Dept. of Interior, U.S. Geological Survey, Washington, DC.

WALKER, B. H. 1992. Biodiversity and ecological redundancy. *Conservation Biology* 6:18–23.

WEBSTER, W. D., J. F. PARNELL, AND W. C. BRIGGS JR. 1985. *Mammals of the Carolinas, Virginia, and Maryland*. University of North Carolina Press, Chapel Hill, North Carolina.

WHITTAKER, R. H. 1956. Vegetation of the Great Smoky Mountains. *Ecological Monographs* 26:1–86.

WIENS, J. A. 1989. Spatial scaling in ecology. *Functional Ecology* 3:385–397.

*WIENS, J. A., AND B. T. MILNE. 1989. Scaling of "landscapes" in landscape ecology, or, landscape ecology from a beetle's perspective. *Landscape Ecology* 3:87–96. Describes research conducted with beetles on "microlandscapes" in grasslands. The fractal structure of microlandscapes affected movement patterns.

*WISER, S. K., R. K. PEET, AND P. S. WHITE. 1998. Prediction of rare-plant occurrence: A southern Appalachian example. *Ecological Applications* 8:909–920. Uses a combination of field-collected data and topographic characteristics to generate predictive equations for four species of rare plants. This analysis sought to understand factors influencing the distribution of these species so that additional populations might be discovered and important habitats protected. The utility of such models is critically examined.

WOFFORD, B. E. 1989. *Guide to the Vascular Plants of the Blue Ridge.* University of Georgia Press, Athens, Georgia.

CHAPTER 14

LANDSCAPE CONTEXT

SCOTT M. PEARSON

OBJECTIVES

Ecological processes at a particular site are the result of both local dynamics and processes acting at the broader scale of the surrounding landscape. Thus, local characteristics and processes are linked to those of broader scales. The potential influence of the surrounding landscape includes effects on community composition, habitat quality, metapopulation dynamics, and local population persistence. Moreover, because landscapes are dynamic, we cannot expect processes and characteristics within habitat patches to remain static if the surrounding landscape is altered by natural or anthropogenic forces. Therefore, understanding the effects of landscape context on habitat patches is an important issue for basic ecological research and for conservation management. The objectives of this exercise are to help students

1. learn to quantitatively describe the "landscape neighborhood" surrounding a specific site;
2. conduct statistical tests to determine if characteristics of the landscape neighborhood are correlated with local community structure; and
3. consider why landscape-level effects may vary among species.

Here, you will determine if the surrounding landscape affects local community structure. Specifically, you will test whether the amount of mature forest in the broader landscape neighborhood affects either the diversity or the occurrence of certain plant species in a patch. In Exercise 1, you will measure the amount of old forest in the landscape surrounding a set of study sites to quantify the landscape neighborhood so that you understand the methods used to gather the data used in Exercise 2. Exercise 2 is an analysis of the effects of landscape context on species diversity using a larger set of data that has been assembled for you. Exercise 3 uses the same data set to test for landscape effects on individual species. The larger data set for Exercises 2 and 3 is available in tabular format in Table 14.3 (a pdf file readable using Adobe Acrobat Reader) in a spreadsheet entitled **context.xls** on the CD.

INTRODUCTION

A growing number of studies have demonstrated that local ecological processes are influenced by patterns and phenomena taking place at broader scales. For example, Gibbs (1998) found that some amphibian species disappeared from forest patches when the percentage of forest in the surrounding landscape dropped below 50%. Similar landscape effects on populations of wintering (Pearson, 1993) and breeding birds (Robinson, 1992) have been documented. The diversity and abundance of wintering birds can depend on the availability of resources in adjacent habitat patches, which are used by local birds for foraging or roosting. Nesting success of some breeding forest birds of North America has been correlated with the rate of brood parasitism by the brown-headed cowbird. The abundance of cowbirds (and resulting parasitism rates) depend on the amount of farmland in the surrounding landscape (Robinson et al., 1995). The spatial pattern of suitable habitat has been shown to affect the local density and sex ratio of small mammals. Collins and Barrett (1997) experimentally fragmented portions of habitat used by meadow voles. In fragmented habitats, the sex ratio was more skewed toward females as compared to unfragmented habitats. Similar effects on population persistence and density exist for plants (Bergelson et al., 1993; Matlack, 1994; Fischer and Stocklin, 1997; Pearson et al., 1998) and insects. The abundance of suitable habitat in the surrounding landscape can affect the persistence of local populations and the chance that a suitable site will be colonized. The surrounding landscape may also influence the probability that a site is disturbed (Turner et al., 1989; O'Neill et al., 1992). Thus, local populations and ecosystem processes exist within the context of the broader landscape.

   In this exercise, you will test hypotheses related to herbaceous species diversity and the landscape context of forested sites supporting these species. The abundance of mature forest surrounding an early successional patch may be important because mature forest may affect the colonization of plant species into early successional patches. This exercise will involve quantifying characteristics of the surrounding landscape from land-cover maps and relating these features to plot-level vegetation data. The specific working hypothesis for the data collection was: *Species richness in young forest is correlated with the proportion of old forest in the surrounding landscape.* Data were collected from a series of plots in early successional forest that were similar with respect to within-site characteristics (e.g., soils, topography, elevation, land-use history) but varied with respect to landscape context. You will test whether the amount of old forest in the landscape affects (1) species richness for the entire plot and (2) the presence/absence of four individual plant species: *Galium aparine*, *Arisaema triphyllum*, *Viola canadensis*, and *Disporum lanuginosum*.

Study Area

The geographic setting is Madison County in western North Carolina, a landscape that has experienced a period of deforestation due to agricultural expansion, followed by a period of reforestation due to the abandonment of agricultural fields. These temperate deciduous forests are characterized by relatively high alpha diversity (i.e., within-patch diversity) in both woody and herbaceous plants. A land-cover map (derived from a 1991 Thematic Mapper satellite image) shows a portion of this landscape (Figure 14.1), including

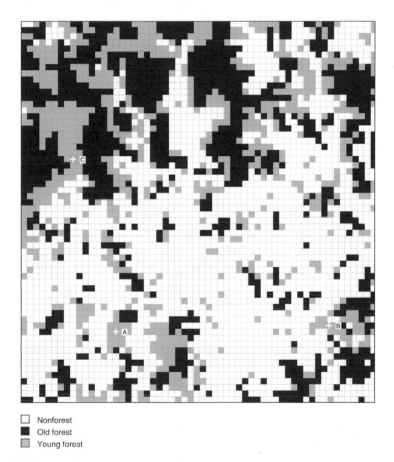

☐ Nonforest
■ Old forest
▨ Young forest

FIGURE 14.1
Land-cover map of study area in Madison County, North Carolina, USA. Each grid cell is 100 × 100 m. Focal cells are marked with a white cross (+) located to the left of the label (A, B, or C). A larger, printable version is available from the CD under the directory for this chapter.

the three primary land-cover types: **nonforest** (mostly agricultural lands with houses), **young forest**, and **old forest**. The young (early successional) forest areas were established after agricultural abandonment around 1950. The old forests were established well before 1950 and show few signs of previous agricultural uses besides woodland grazing.

EXERCISES

You will use data based on a field study designed to test for the influence of landscape context on herbaceous species occurring in patches of young forest. Using the land-cover map (Figure 14.1), half-hectare study plots were randomly located within patches of young forest at least 200 meters apart. The vegetation within each plot was sampled. In addition, a land-cover map was used to determine the proportion of old forest in an 81-hectare square window (9 × 9 cell area centered on the study plot).

In Exercise 1, you will collect landscape context data and calculate species richness for three practice sites to ensure that you understand how these data were collected. Then, in Exercise 2, you will conduct a more extensive analysis using data from 80 plots. While Exercise 2 focuses on plant species diversity, Exercise 3 will address the effect of landscape context on selected species.

EXERCISE 1
Data Collection

LANDSCAPE CONTEXT DATA

Land-cover maps can be used to characterize the landscape surrounding individual sample points. For this study, we are interested in the percentage of a 81-hectare "landscape neighborhood" that is composed of old forest. These old forests serve as a reservoir of species that may colonize a patch of young forest as it develops. The procedure for quantifying landscape context used here is identical to that used for the larger data set you will analyze in Exercises 2 and 3.

Using the land-cover map (Figure 14.1), measure the portion of old forest surrounding study plots A, B, and C as follows. Note that the location of each study plot is identified by a white cross (+), and to the right of the actual plot is a plot label (A, B, or C). You may wish to use the larger version of Figure 14.1 which can be printed from the CD.

1. Sketch an 81-hectare window (9 × 9 cells) on the map to delineate the neighborhood around each study plot. Center the window on the focal cell indicated with a white cross (to the left of the label).

2. Count the number of cells of old forest in the window and record your findings in Table 14.1 which can be viewed using Adobe Acrobat Reader and then printed from the CD.

3. Calculate the percent of old forest in the window surrounding each plot. Again, record your results in Table 14.1.

PLOT-LEVEL VEGETATION DATA

The study plots were also surveyed in the field, and a species list of herbaceous plants was compiled for each plot. Examine the species lists for plots A, B, and C in Table 14.2 on the CD. Tally up the total number of species per plot and enter the count on the last line of Table 14.2.

Question 1. Which plot has the greatest species richness?

Question 2. Look at the land-cover maps and describe the pattern of land cover in the vicinity of each of the three plots.

Question 3. From this small sample, does it appear that species diversity is related to the abundance of old forest in this landscape?

EXERCISE 2
Effects of Landscape Context on Species Richness

Next, you'll determine if the abundance of old forest in the surrounding landscape affects local species richness using a much larger data set (Table 14.3 or **context.xls** on the CD). You can analyze the data "by hand" using the table, or using a spreadsheet. Note that the landscape context measurements have already been determined for you.

1. To streamline the analysis, classify the plots into four landscape context categories based on the percentage of old forest in the surrounding landscape: (a) <10% old forest, (b) 10–39%, (c) 40–60%, and (d) >60%.

2. Calculate the mean ($\bar{x}$) species richness and standard deviation (s) for each of the four landscape context categories using the following formulas:

$$\bar{x} = \frac{\sum x_i}{n} \qquad s = \sqrt{\frac{\sum (\bar{x} - x_i)^2}{n - 1}}$$

where x_i is an individual datum for a plot and n is the number of data points (i.e., 20 in this case). Enter your results in Table 14.4 which can be printed from the CD.

3. Calculate a 95% confidence interval for each of these means using the following formula: $95\% \ CI = \bar{x} \pm \frac{1.96s}{\sqrt{20}}$. Enter your results in Table 14.4.

4. For each possible pair of landscape context categories (e.g., <10% and 10–39%), determine whether the confidence intervals overlap. Nonoverlapping confidence intervals indicate that species richness dif-

fers between the two categories of landscape context; overlapping confidence intervals indicate that the mean species richness values are not significantly different at the $P \leq 0.05$ level.

Question 4. Prepare a bar graph on a separate sheet of paper plotting *mean species richness* (y axis) against *old forest in landscape* (x-axis). Summarize the trends evident in the table and bar graph.

EXERCISE 3
Influence of Landscape Context on the Presence of Individual Species

Next (using Table 14.3 from the CD), you'll determine the importance of landscape context for several individual species: *G. aparine*, *A. triphyllum*, *V. canadensis*, and *D. lanuginosum*.

Galium aparine (cleavers) is a sprawling, prickly plant found in many different forest and nonforest habitats (Newcomb, 1977). Fruits, stems, and leaves have small hooks that attach to fur and feathers allowing propagules to be transported by vertebrates. It is able to grow in a wide variety of forested habitats but is most abundant in moist, fertile areas.

Arisaema triphyllum (jack-in-the-pulpit) is widespread in young and old forested habitats. It produces red berries eaten and dispersed by birds and mammals. A single plant may live over five years, and individuals store energy in an underground corm. Each year an individual produces a single flower that is either male or female. Moreover, individuals are able to change sex from year to year (Doust et al., 1986).

Viola canadensis (Canada violet) and *Disporum lanuginosum* (yellow mandarin) are found in mesic woods (Weakley, 1995). They seem to be sensitive to forest disturbance because they are most abundant in the old forest (least disturbed) sites. The violet is more widespread and over ten times more abundant than the mandarin. Both species are dispersed by ants.

Based on the life-history information presented above, classify each species according to habitat needs (specialist or generalist) and dispersal ability (good or poor). Enter your results in Table 14.5 from the CD. Before conducting any analyses, using Table 14.5 and the information on each species' life history, hypothesize about the effects of landscape context on each of these species. That is, hypothesize whether you think each species should be negatively or positively correlated with the percent of the surrounding landscape neighborhood occupied by old forest.

Next, you will use contingency table analysis to determine whether the presence/absence of specific species is affected by landscape context.

1. Consider the species *Galium aparine*. Using the data in Table 14.3 from the CD, count the total number of plots in which this species is present in plots with < 10% old forest in the landscape neighbor-

hood. This constitutes the **"observed"** value. Record this number in the *Galium aparine* table found in Table 14.6, in the row labeled "Present," then "Observed," under the column labeled "<10%." Repeat for the other landscape context categories.

2. Similarly, determine the number of plots in which this species is absent. Record your counts in the Table 14.6 in the row labeled **"Absent,"** then **"Observed."**

3. Calculate the row and column totals for the **observed** data. Add the two row totals; this sum is called the **grand total.**

4. Next, calculate the **expected** value for each landscape context category, for present and absent (8 calculation), using the following equation:

$$expected = \frac{(row\ total) * (column\ total)}{grand\ total}$$

Enter your results in the table in the rows labeled "Expected" for the appropriate categories.

5. A chi-square (χ^2) statistic can be used to test for a statistically significant influence of landscape context on the presence or absence of each species. This type of statistic is frequently used in contingency table analyses. Calculate the chi-square value as follows:

$\chi^2 = \sum \frac{(observed - expected)^2}{expected}$. Enter your results in Table 14.7 from the CD.

6. If the χ^2 value exceeds 7.81, presence is significantly dependent on landscape context at the $P < 0.05$ level. This test has degrees of freedom = 3. Ask your instructor for assistance if this type of statistical test is unfamiliar to you. Determine whether the χ^2 test for *Galium aparine* is significant or not significant and record in Table 14.7.

7. Repeat steps 1–6 for the other three species.

Question 5. Describe the results for the four species based on the numbers in the contingency tables and the statistical analysis.

Question 6. How do these results compare with your initial hypotheses?

DISCUSSION QUESTIONS

1. According to your results in Exercise 2, is species richness correlated with landscape context at the scale used in this analysis?

2. List some ecological processes that might have produced the patterns observed in Exercise 2. These proposed mechanisms can serve as hypotheses to be tested in further field studies.

3. Do the patterns of presence/absence shown by the four species seem to be affected by landscape context? If yes, propose some plausible mechanisms. How does the natural history of each species relate to its pattern of presence/absence?

4. Would changing the size of the "landscape window" (e.g., to more than or less than 81 ha) affect the findings? How would using (a) a much smaller and (b) a much larger window affect your ability to detect an effect of landscape context?

5. This exercise focused on the effect of surrounding old forest on herbaceous plant communities. What other land covers or landscape elements might affect plant or animal communities? List some other landscape elements and their hypothesized effects.

6. Pick one of your hypothesized elements from Discussion Question 5 and design a study to test for its effects.

7. Can you think of a situation in which landscape context would affect an ecosystem process (e.g., nutrient cycling) within a patch?

BIBLIOGRAPHY

Note. An asterisk preceding the entry indicates that it is a suggested reading.

*ANDREN, H. 1992. Corvid density and nest predation in relation to forest fragmentation: A landscape perspective. *Ecology* 73:794–804. Examines the threat of predation on the nests of songbirds. The abundance of corvid predators and predation rates on nests were affected by landscape-level abundance of agricultural versus forest habitats.

BERGELSON, J., J. A. NEWMAN, AND E. M. FLORESROUX. 1993. Rates of weed spread in spatially heterogeneous environments. *Ecology* 74:999–1011.

COLLINS, R. J., AND G. W. BARRETT. 1997. Effects of habitat fragmentation on meadow vole (*Microtus pennsylvanicus*) population dynamics in experimental landscape patches. *Landscape Ecology* 12:63–76.

DOUST, L. L., J. L. DOUST, AND K. TURI. 1986. Fecundity and size relationships in jack-in-the-pulpit, *Arisaema triphyllum* (Araceae). *American Journal of Botany* 73:489–494.

*DUNNING, J. B., B. J. DANIELSON, AND H. R. PULLIAM. 1992. Ecological processes that affect populations in complex landscapes. *Oikos* 65:165–179. Summarizes the effects of the juxtaposition and spatial arrangement of habitats on population dynamics, drawing on both theory and empirical studies.

FISCHER, M., AND J. STOCKLIN. 1997. Local extinctions of plants in remnants of extensively used calcareous grasslands 1950–1985. *Conservation Biology* 11:727–737.

GIBBS, J.P. 1998. Amphibian movements in response to forest edges, roads, and streambeds in southern New England. *Journal of Wildlife Management* 62:584–589.

*GLENN, S. M., AND S. L. COLLINS. 1992. Effects of scale and disturbance on rates of immigration and extinction of species in prairies. *Oikos* 63:273–280. Discusses spatial and temporal dynamics of plant communities in prairies. Illustrates how local plant communities are linked to regional species pool and are influenced by ecosystem processes such as disturbance.

*MATLACK, G. R. 1994. Vegetation dynamics of the forest edge–trends in space and successional time. *Journal of Ecology* 82:113–123. Describes succession in plant community under the influence of adjacent habitats.

*MCCOLLOUGH, D. R., ED. 1996. *Metapopulations and Wildlife Conservation.* Island Press, Washington, DC. Examines a wide range of topics related to metapopulations, an important concept for understanding populations and communities at landscape scales.

NEWCOMB, L. 1977. *Newcomb's Wildflower Guide.* Little, Brown, and Company, Boston.

O'NEILL, R. V., R. H. GARDNER, M. G. TURNER, AND W. H. ROMME. 1992. Epidemiology theory and disturbance spread on landscapes. *Landscape Ecology* 7:19–26.

*PEARSON, S. M. 1993. The spatial extent and relative influence of landscape-level factors in wintering bird populations. *Landscape Ecology* 8:3–18. Compares the relative effects of within-patch versus landscape habitats. Species responded in different ways to landscape context of focal patches.

*PEARSON, S. M., A. B. SMITH, AND M. G. TURNER. 1998. Forest fragmentation, land use, and cove-forest herbs in the French Broad River Basin. *Castanea* 63:382–394. Documents that forest patch size affects species richness and abundance in herb communities. Patch-size effects depend on differences in the natural history of species.

ROBINSON, S. K. 1992. Population dynamics of breeding Neotropical migrants in a fragmented Illinois landscape. In *Ecology and Conservation of Neotropical Migrant Landbirds,* J. M. Hagan and D. W. Johnson, eds. Smithsonian Institution Press, Washington, DC, pp. 408–418.

*ROBINSON, S. K., F. R. ROBINSON III, T. M. DONOVAN, D. WHITEHEAD, AND J. FAABORG. 1995. Regional forest fragmentation and the nesting success of migratory birds. *Science* 267:1987–1990. Reviews the breeding success of migrant birds that require forest in landscapes dominated by agriculture.

TURNER, M. G., R. H. GARDNER, V. H. DALE, AND R. V. O'NEILL. 1989. Predicting the spread of disturbance across heterogeneous landscapes. *Oikos* 55:121–129.

WEAKLEY, A. S. 1995. *Flora of the Carolinas and Virginia* (working draft). The Nature Conservancy, Southeast Regional Office, 101 Conner Drive, Chapel Hill, NC 27514.

CHAPTER

15

LANDSCAPE CONNECTIVITY AND METAPOPULATION DYNAMICS

KIMBERLY A. WITH

OBJECTIVES

The wholesale destruction and fragmentation of habitat by humans has precipitated a global extinction crisis in which species are going extinct faster than at any other time since the last mass extinction event occurred some 65 million years ago (Wilcox and Murphy, 1985). Besides resulting in an absolute loss of available habitat, habitat destruction may also fragment the landscape. The process of **habitat fragmentation** involves the dissection of remaining habitat into a greater number of smaller and increasingly more isolated patches. Habitat fragmentation disrupts landscape connectivity, which may interfere with dispersal, leading to more isolated populations, and thereby enhancing extinction risk for species. Although habitat destruction may ultimately have a greater effect on populations than fragmentation (Fahrig, 1997), habitat fragmentation may exacerbate or hasten the effects of habitat loss. There has thus been considerable interest within the field of conservation biology in predicting the consequences of habitat loss and fragmentation for species persistence.

This lab explores how quantitative methods and theory derived from landscape ecology can be used to predict the consequences of landscape connectivity for metapopulation persistence. This lab will enable students to

1. develop species' perceptions of landscape connectivity by using neutral landscape models to quantify critical thresholds of habitat fragmentation for species with different gap-crossing abilities (Part 1);

2. distinguish between structural and functional connectivity of landscape structure (Part 1);
3. explore the consequences of landscape connectivity for metapopulations by identifying extinction thresholds for species with different demographic potentials in landscapes subjected to various levels of habitat loss and fragmentation (Part 2); and
4. assess what types of species are most sensitive to habitat loss and fragmentation (Part 2).

The lab is divided into two sections, which can be done independently or in sequence. The first section, **Part 1, Species' Perceptions of Landscape Connectivity,** is a full lab that will enable you to develop species-centered definitions of landscape connectivity using the computer program RULE (see **Chapter 9, Neutral Landscape Models**). You will generate landscapes across a range of habitat loss and fragmentation and then determine whether these landscapes are connected from the standpoint of species with different dispersal abilities. The second section, **Part 2, Extinction Thresholds for Species in Fragmented Landscapes,** is a graphical analysis of species' persistence and extinction risk in fragmented landscapes, resulting from a recent modeling synthesis between neutral landscape models and metapopulation theory (With and King, 1999a). This shorter section can also be used in a discussion session.

The required materials for Part 1 are:
1. Computer running Microsoft Windows
2. Copy of RULE software and familiarity running RULE (see **Chapter 9, Neutral Landscape Models**)
3. Set of batch files (4) and associated script files (9 script files per batch file = 36 script files) to permit automation of RULE for these exercises
4. Spreadsheet software (e.g., Excel)
5. Tables 15.1 and 15.2 which can be printed from the CD using Adobe Acrobat Reader (which is freely available on the web).

Part 2 requires:
1. Familiarity with the paper by Lande (1987) on extinction thresholds in demographic models of territorial populations
2. Graphical output from spatially explicit demographic model depicting extinction thresholds for different species on fractal landscapes (Figures 15.5–15.8). The CD contains larger, printable versions of these figures.

PART 1. SPECIES' PERCEPTIONS OF LANDSCAPE CONNECTIVITY

INTRODUCTION

Landscape connectivity, a central theme in both landscape ecology and conservation biology (With, 1997), is also a vital element of landscape structure

(Taylor et al., 1993). **Landscape connectivity** has been defined as "the functional relationship among habitat patches, owing to the spatial contagion of habitat and the movement responses of organisms to landscape structure" (With et al., 1997: 151). Thus, landscape connectivity is not just a function of the spatial arrangement of habitat, but also depends on a species' habitat affinities (habitat preferences or associations) and the ability of individuals to move through elements of the landscape. Chapter 13, Interpreting Landscape Patterns from Organism-Based Perspectives, demonstrated how a landscape may appear different, in terms of its patch structure and overall connectivity, to species with different habitat requirements. Beyond a species' habitat affinities, certain landscape elements, such as corridors, may facilitate movement, whereas other elements may impede movement because they represent inhospitable environments or are structurally "viscous" habitats that are difficult to traverse (With and Crist, 1995; Gustafson and Gardner, 1996; With et al., 1997). Furthermore, the dispersal or movement abilities of species also affect whether the landscape is perceived as connected. Species with good dispersal abilities or "gap-crossing abilities" can traverse areas of unsuitable habitat and are likely to perceive the landscape as connected across a greater range of habitat loss than are species with poor dispersal abilities (Dale et al., 1994; With and Crist, 1995; Pearson et al., 1996).

In landscape ecology, neutral landscape models have been used to quantify landscape connectivity (Gardner et al., 1987; Gardner and O'Neill, 1991; With and King, 1997; With, 1997). In this context, landscape connectivity is defined by the **percolation threshold,** the amount of habitat at which the landscape is no longer connected by a continuous cluster of habitat that spans the entire landscape (Figure 15.1). Above this threshold, habitat destruction merely results in a total loss of habitat area; that is, the change in landscape structure is a quantitative one (Andrén, 1994). Below the threshold, however, habitat loss produces a qualitative change in landscape structure, in which the landscape becomes fragmented into numerous small, isolated habitat patches. Further habitat loss below the threshold again produces a quantitative change in landscape structure, in terms of an increase in fragmentation.

Disruption of landscape connectivity is thus predicted to occur as a threshold phenomenon in which a small loss of habitat at the critical threshold abruptly disconnects the landscape. The use of neutral landscape models to define critical thresholds in landscape connectivity was explored in **Chapter 9, Neutral Landscape Models.** The critical threshold at which landscapes become disconnected depends on the underlying habitat distribution (random or fractal; fractal algorithms are used to create spatially correlated habitat distributions) and the rule of habitat connectivity (e.g., four-cell or eight-cell neighborhood rule). These rules of habitat connectivity can also be equated to the movement or gap-crossing abilities of a species (Pearson et al., 1996). The most restrictive movement rule constrains the organism to move only within suitable habitat cells; that is, the species is unable to cross gaps of unsuitable habitat. This movement pattern corresponds with the nearest-neighbor (four-cell) rule of habitat connectivity (rule 1, Figure 15.2). Alternatively,

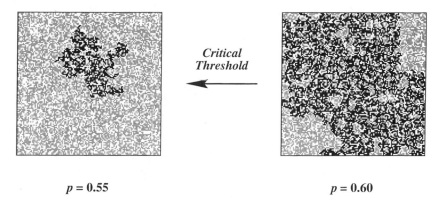

p = 0.55   *p = 0.60*

FIGURE 15.1
Landscape connectivity is defined by the presence of a *percolation cluster*, a continuous cluster of habitat that spans the entire landscape (black cells, $p = 0.60$, see map on right). Habitat fragmentation, a disruption in landscape connectivity, occurs abruptly as a threshold phenomenon. A small loss of habitat near the critical threshold fragments the percolation cluster, thereby disrupting landscape connectivity. Note that the largest cluster of habitat (shaded in black) on the map with 55% habitat ($p = 0.55$) no longer spans the map in this random neutral landscape (patch connectivity defined by four-cell neighborhood or rule 1).

a species capable of crossing gaps (one-cell wide) of unsuitable habitat could be modeled using a 12-cell neighborhood rule (rule 3, Figure 15.2).

Landscape connectivity for species with different dispersal abilities can thus be assayed using different movement rules. This also permits a comparison between the **structural connectivity** of landscapes, which is the physical adjacency of habitat cells as defined by the nearest-neighbor (four-cell) rule, and the **functional connectivity** of landscapes, in which patches are connected by the movement or dispersal abilities of species. In this lab exercise, you will quantify the connectivity of landscapes from the perspective of species with different gap-crossing abilities.

Recall that landscape connectivity is assessed by how an organism interacts with the spatial distribution of habitat, in terms of a species' habitat affinities and individual movement responses to landscape structure. In heterogeneous landscapes, assessment of landscape connectivity is complicated if a species uses several habitat types to varying degrees and its movement is differentially affected by habitat structure (e.g., With and Crist, 1995, With et al., 1997).

To keep things simple, you will generate binary (habitat vs. nonhabitat) landscapes in this exercise, in which each species has the same habitat requirement (e.g., forest), but differs in the ability to traverse unsuitable habitat (e.g., agricultural fields). The **poor disperser** is constrained to move only through adjacent habitat cells (four-cell neighborhood, rule 1) and cannot cross gaps of unsuitable habitat. An example of such a species might be a

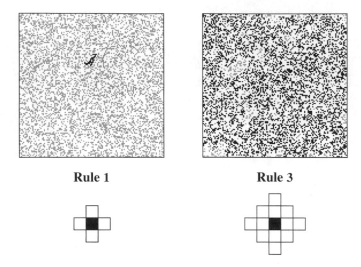

Rule 1   **Rule 3**

FIGURE 15.2
Whether a landscape is connected depends on the rule used to define patch connectivity. Each of these two landscapes has 35% habitat. The largest patch on each landscape is highlighted in black. If patch connectivity is assessed using the nearest-neighbor rule (rule 1 or four-cell neighborhood), then the landscape is not connected. If patch connectivity is assessed using rule 3 (12-cell neighborhood), perhaps corresponding to a species with gap-crossing abilities, then this random neutral landscape is connected owing to the presence of a percolating cluster that spans the entire landscape. (Modified from With, 1997.)

woodland salamander (*Plethodon* spp.), which is susceptible to desiccation if it ventures beyond the moist forest floor, or a white-footed mouse (*Peromyscus leucopus*), which lives in forest fragments and rarely ventures across agricultural fields (except during juvenile dispersal) where it is more vulnerable to predators. The **good disperser** is capable of crossing gaps (one-cell wide) of unsuitable habitat (12-cell neighborhood, rule 3). For example, many birds species (e.g., blue jay, *Cyanocitta cristata*) readily fly across agricultural fields in their movements among woodlots.

Neutral Landscape Models

As a basis for evaluating how habitat loss and fragmentation affect landscape connectivity, you will generate fractal landscape patterns, which have the advantage of permitting an exploration of the relative effects of habitat loss (p) apart from fragmentation (the relative degree of habitat clumping or degree of spatial contagion, H) on species' perceptions of landscape connectivity. Fractal landscapes generated with negative spatial autocorrelation ($H = 0.0$) appear more fragmented than fractal landscapes with positive spatial autocorrelation ($H = 1.0$; Figure 15.3). Fragmented fractal landscapes ($H = 0.0$) have more small patches and more edge than clumped fractal landscapes ($H =$

Random   *H* = 0.0   *H* = 0.5   *H* = 1.0

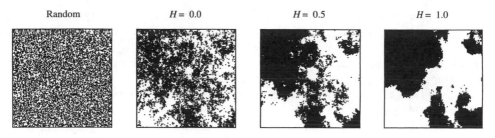

FIGURE 15.3

Examples of random and fractal neutral landscapes. All landscapes contain 50% habitat (black cells). Spatial contagion (*H*) can be adjusted in fractal landscapes to produce landscapes that vary in degree of habitat fragmentation, from clumped (*H* = 1.0) to extremely fragmented (*H* = 0.0).

1.0; With and King, 1999b). From a management standpoint, the fragmented fractal landscapes might represent a scenario in which habitat destruction occurs at a fine spatial scale (e.g., selective logging), whereas the clumped fractal landscapes represent habitat destruction at a broader or coarser scale (e.g., extensive clear-cutting). The designation of these fractal landscapes as "fragmented" or "clumped" is done for convenience only; assessment of whether these landscapes are indeed connected will ultimately depend on species' perceptions of landscape structure (i.e., gap-crossing abilities and the functional connectivity of habitat).

Analysis of Landscape Connectivity using RULE

Because it is assumed that you have experience running RULE interactively, batch files and their associated script files have been provided for the exercises in this lab to facilitate computer runs and to enable you to complete these exercises in a reasonable time frame. The batch files (*.bat) permit sequential runs of RULE to be performed automatically. The information for each run is listed in a separate script file (*.scr), which contains the dialog that you would have to enter if you were running RULE interactively. There are four batch files (one for each exercise), each of which has nine script files associated with it (36 files total). *These files must be loaded into the same directory where RULE resides* (e.g., C:\RULE\). Refer to Chapter 9, Neutral Landscape Models, for additional information on how to run RULE in batch mode.

Each of the four batch files (*.bat) pertains to a separate exercise involving either a poor or good disperser (rule 1 or rule 3, respectively) on either a clumped or fragmented (*H* = 1.0 or *H* = 0.0, respectively) fractal landscape (Figure 15.3). Thus, poorH1.bat is the batch file for poor dispersers on clumped fractal landscapes, and goodH0.bat is the batch file for good dispersers on fragmented fractal landscapes. Each script file specifies the map type (*m* for multifractal, hereafter referred to simply as "fractal"), the map size (7 levels = 128 × 128 cells), the degree of clumping or spatial contagion (*H* = 0.0 or 1.0), the movement or neighborhood rule by which patch connectivity is assessed (rule 1 or rule 3), the number of habitat types (two—habitat vs. nonhabitat), the relative abundance of each habitat type (e.g., p_1 = 0.1, p_2 = 0.9), and the number of replicate maps generated for analysis (n = 10). Consider splitting the exercises among individuals within a group, or among different groups in the class, to facilitate run time.

NOTE: There is a slight difference in the script files between this lab and Chapter 9, Neutral Landscape Models. Chapter 9 uses the 1st probability to define habitat 1, and the 0th probability to define nonhabitat. In contrast, this exercise assumes the 0th probability to be 0 (nonexistent) and uses the 1st probability (habitat 1) to define habitat and the 2nd probability (habitat 2) to define nonhabitat. This difference in how habitat is assigned has no effect on the landscape maps generated by RULE.

Determining Thresholds in Landscape Connectivity

Of the various patch metrics listed in the output file (rulerun.log) generated by RULE, you will only be interested in **percolation frequency**, which will be used to quantify landscape connectivity from the perspective of different species in these various landscape scenarios. The use of spreadsheet software (e.g., Excel) to graph these data (p vs. percolation frequency) is therefore recommended; otherwise, these data can be plotted manually on graph paper. The percolation threshold (p_{crit}) is defined as the level of habitat abundance (p) at which the probability of having a connected landscape (defined by percolation frequency) is less than or equal to 0.5. Above this value, the landscape is considered to be connected; below the threshold, the landscape is considered to be disconnected. Draw a horizontal line on your graph from where the percolation frequency is 0.5 on the y-axis to where it intersects the response curve (i.e., the line of data you plotted). At this point of intersection, draw a vertical line down to the x-axis. This is the level of habitat abundance (p) at which there is a 50% probability of having a connected landscape (i.e., the percolation threshold, p_{crit}).

EXERCISES

Before starting these exercises, generate some *a priori* hypotheses in Table 15.1 (which can be printed from the CD) about how thresholds in landscape connectivity will be affected by dispersal ability and landscape structure (e.g., how do you expect p_{crit} for poor dispersers on fragmented fractal landscapes to compare with that of good dispersers on these same landscapes?).

EXERCISE 1.1
Poor Disperses in Clumped Landscapes

1. Assuming that you are working in a Windows environment, open an MS-DOS window, move to the directory where RULE is located (and where the batch and script files for this exercise must also be located),

and type **"poorH1.bat"** at the prompt. Example: **C:\RULE > poorH1.bat.** Alternatively, you can double-click on the icon for this batch file within the File Manager, which causes an MS-DOS window to open and RULE to be executed automatically.

2. Each of the nine script files (pH1p*.scr) associated with this batch file is run sequentially. There are nine levels of habitat abundance at which landscape connectivity is being quantified, and thus there are nine script files. Note that it will take some time for all nine runs to be completed, so be patient! What are the nine levels of habitat abundance (*p*) being generated across these landscapes?

3. After each run, RULE creates an output file (rulerun.log) with the various patch metrics (including *percolation frequency*) from the analysis of the landscape maps you generated. The batch file copies each output file to a new file with the same naming convention as the script files (e.g., pH1p10.log). Thus, when the batch file has completed its runs, you should have nine output files, one for each script file.

4. Print out or open each output file (pH1p*.log) and record the percolation frequency in **Table 15.2** from the CD for **poor dispersers** and **clumped landscapes.** Note that each output file corresponds to one row of data in the table.

5. Enter the data for habitat abundance (*p*) and percolation frequency in a spreadsheet and plot as a line graph. Your graph should have habitat abundance (*p*) on the abscissa (*x*-axis) and percolation frequency on the ordinate (*y*-axis).

6. Identify the critical threshold (*p_{crit}*) of landscape connectivity (see **Part 1, Determining Thresholds in Landscape Connectivity**).

Question 1.1. What is the movement rule for dispersers in this exercise?

Question 1.2. What is the spatial contagion (*H*) of the fractal landscapes in this exercise?

Question 1.3. What is the threshold of landscape connectivity (*p_{crit}*) for poor dispersers in clumped fractal landscapes?

EXERCISE 1.2
Good Dispersers in Clumped Landscapes

1. Repeat steps 1–3 in Exercise 1.1 but execute the batch file **goodH1.bat** and run its associated script files (gH1p*.scr). The output files generated at the completion of the runs follow the naming convention of the script files (e.g., gH1p*.log).

2. Record the data for percolation frequency in **Table 15.2** for **good dispersers** and **clumped landscapes.**

3. Enter the data for habitat abundance (*p*) and percolation frequency into a spreadsheet and plot these data as a line graph. Note that you

can enter the percolation frequency data for good dispersers in the same spreadsheet that you used for poor dispersers and create a graph that displays both curves. This will facilitate comparisons of how good and poor dispersers perceive landscape connectivity in clumped landscapes.

4. Assess the landscape connectivity threshold (p_{crit}) for good dispersers.

Question 1.4. What is the movement rule for dispersers in this exercise?

Question 1.5. What is the threshold (p_{crit}) at which good dispersers perceive the clumped fractal landscape to be disconnected?

EXERCISE 1.3
Poor Dispersers in Fragmented Landscapes

1. Follow steps 1–3 from Exercise 1.1, only this time execute the batch file **poorH0.bat** and its associated script files (pH0p*.scr), which will generate separate output files reporting the patch metrics (pH0p*.log).

2. Record the data for percolation frequency in **Table 15.2** on the CD for **poor dispersers** in **fragmented landscapes.**

3. Enter the data for habitat abundance (p) and percolation frequency into a spreadsheet and plot as a line graph. You may wish to do this as a separate graph initially, but you can also plot the curves for poor dispersers on a single graph to facilitate comparisons between clumped and fragmented landscapes.

4. Evaluate the landscape connectivity threshold (p_{crit}) for poor dispersers on fragmented landscapes.

Question 1.6. What is the spatial contagion (H) of fragmented fractal landscapes?

Question 1.7. At what level of habitat abundance (p) do poor dispersers perceive fragmented fractal landscapes to become disconnected?

EXERCISE 1.4
Good Dispersers in Fragmented Landscapes

1. Follow steps 1–3 from Exercise 1.1, but execute batch file **goodH0.bat,** which will run the script files gH0p*.scr and generate the output files gH0p*.log.

2. Record the data for percolation frequency in **Table 15.2** for **good dispersers in fragmented landscapes.**

3. Enter the data for habitat abundance (p) and percolation frequency in a spreadsheet and create a line graph. Again, you may initially wish to plot these data separately, but you can then combine these data with that for good dispersers on clumped landscapes to evaluate how

good dispersers perceive landscape connectivity in fragmented and clumped landscapes.

4. Evaluate the landscape connectivity threshold (p_{crit}) for good dispersers in fragmented landscapes.

Question 1.8. What is the connectivity threshold (p_{crit}) for good dispersers in fragmented fractal landscapes?

DISCUSSION QUESTIONS

Answer these questions on additional sheets of paper or discuss as a group.

1. Compare your initial expectations of how thresholds in landscape connectivity would be affected by species' dispersal ability and landscape structure (Table 15.1) with the actual thresholds determined from your analyses. Are there any counter-intuitive results? Explain.
2. Are good dispersers more sensitive, or less sensitive to habitat loss than poor dispersers? Compare the value of the landscape connectivity threshold (p_{crit}) for good and poor dispersers within each type of landscape (e.g., clumped fractal landscape). Do good dispersers perceive the landscape to be disconnected sooner (threshold occurs at higher values of p) or later (threshold occurs at lower values of p) than poor dispersers?
3. How does the scale of habitat removal affect a species' perception of landscape connectivity? Compare the landscape connectivity threshold (p_{crit}) in the clumped fractal landscape (coarse-scale habitat destruction) with the threshold for that disperser in the fragmented fractal landscapes (fine-scale habitat destruction) for each disperser.
4. How does changing the scale of habitat removal differentially affect poor versus good dispersers? Is the shift in the connectivity threshold (p_{crit}) between clumped and fragmented fractal landscapes greatest for poor dispersers or for good dispersers?
5. Suppose that 45% of the habitat on the landscape is to be destroyed. What would you advise regarding the scale of habitat removal so as to maintain landscape connectivity for both poor and good dispersers? Would it be better to remove habitat in numerous small cells (fragmented fractal landscape pattern) or to clear-cut a few large areas (clumped fractal landscape pattern)? Explain.
6. If habitat can be functionally connected by dispersal, what does this imply about the need for habitat corridors to enhance landscape connectivity?
7. Is species persistence guaranteed on a connected landscape? In other words, is landscape connectivity necessary and sufficient for population persistence? What factors other than dispersal ability might affect whether a species is able to persist on a landscape?

PART 2. EXTINCTION THRESHOLDS FOR SPECIES IN FRAGMENTED LANDSCAPES

INTRODUCTION

A disruption in landscape connectivity should have serious consequences for population dynamics and species persistence in fragmented landscapes (With,

1997; Wiens, 1997a). To what extent does habitat fragmentation result in a disruption of metapopulation dynamics and an increase in extinction risk for species with different life-history traits and dispersal abilities? A **metapopulation** is a collection of spatially subdivided populations that are connected by dispersal. Metapopulation structure enhances the regional persistence of the entire population; although individual populations go extinct locally, dispersers from nearby populations eventually may be able to recolonize habitat patches. Dispersal is the "glue" that keeps the metapopulation together (Hansson, 1991). Too much dispersal, however, would bring the individual populations into synch, such that if one population went extinct, they all would go extinct. Too little dispersal would isolate populations and prevent recolonization of extinct patches. Thus, there is a critical level of dispersal and geometry of patches (i.e., connectivity) that gives rise to metapopulation dynamics, which has been whimsically referred to as the "Goldilock's Zone" by John Wiens (1997b).

Extinction Thresholds in Demographic Models

Russell Lande (1987) derived a general demographic model from Levins' (1969) classic metapopulation model to predict extinction thresholds for territorial vertebrates (e.g., birds) that differed in their demographic potential (a combination of reproductive output and dispersal ability). **Extinction thresholds** are defined as the amount of habitat at which the proportion of suitable habitat patches occupied by the population at equilibrium, p^*, equals 0. The derivation of this demographic model is outlined in Lande (1987). For the purposes of this lab, the following parameters will be useful to know.

MODEL INPUT

h, amount of suitable habitat. Note that this is equivalent to habitat abundance (p) that we have been using in the first part of this lab. In this metapopulation model, the parameter p has a different meaning (see below).

p, proportion of habitat patches occupied by the species in the landscape.

R_0', net lifetime reproductive output. The number of female offspring produced per female, contingent on the female having found a suitable territory.

m, number of sites (cells) a disperser can visit in its search for a suitable, unoccupied territory before it perishes. This can be equated to the dispersal ability of the species.

MODEL OUTPUT

p^*, proportion of habitat patches occupied by the species at demographic equilibrium (Table 15.3).

k, demographic potential of the population (Table 15.3). This is the maximum proportion of habitat patches that can be occupied at equilibrium (p^*) when the entire landscape is suitable ($p^* = k$ when $h = 1.0$). If it is assumed that the population is at demographic equilibrium (how realistic is this for a

TABLE 15.3
COMPARISON OF MODEL STRUCTURE BETWEEN LANDE'S (1987)
DEMOGRAPHIC MODEL AND WITH AND KING'S (1999) MODELING
SYNTHESIS OF METAPOPULATION AND FRACTAL NEUTRAL
LANDSCAPE MODELS.

Model Terms	Lande Model[1]	With-King Model[2]
Dispersal success (probability juvenile finds suitable habitat in m attempts)	$1 - (1 - \epsilon)(u + ph)^m$	$1 - (1 - \epsilon)[(1 - h')^{m^{\beta_1}} + (ph')^{m^{\beta_2}}]$
Equilibrium patch occupancy (p^*)	$1 - (1 - k)/h$	$[(k' - (1 - h')^{m^{\beta_1}})^{m^{\frac{1}{\beta_2}}}]/h$
Demographic potential (k)	$[(1 - 1/R_0')/(1 - \epsilon)]^{1/m}$	$[(1 - 1/R_0')/(1 - \epsilon)]$

[1]Parameters: ϵ = probability juvenile inherits natal territory ($\epsilon = 0$ in With-King model; i.e., obligate dispersal); p = proportion of patches initially occupied; h = amount of suitable habitat ($1 - h$); m = number of dispersal steps in search for suitable habitat; R_0' = net lifetime reproductive output (female offspring per female).
[2]In the With-King model, dispersal success was derived as a fitted function on simulated dispersal success on fractal landscapes. Thus, $h' = a + bh$, where h is the abundance of habitat and a and b are fitted parameters that vary with the spatial contagion H of the landscape. The parameters β_1 and β_2 are also fitted; β_1 varies with H, and β_2 varies with both H and h.

population on a landscape that has recently undergone habitat loss and fragmentation?), then k can be estimated directly from p^* and h in Lande's model, without more detailed information on life history (R_0') and dispersal abilities (m) of the species, as $k = 1 - (1 - p^*) h$ (see formula for p^* in Lande's model, Table 15.3).

h_{crit}, the extinction threshold. This is the level of habitat abundance at which $p^* = 0$. In Lande's model, which assumes a random distribution of suitable habitat, the extinction threshold is defined simply as $h_{crit} = 1 - k$.

Model Predictions

This model generated several interesting predictions (Figure 15.4):

1. Not all habitat patches are occupied, even when the entire landscape is suitable ($h = 1.0$). Thus, a species with a demographic potential of $k = 0.95$ would only be able to occupy 95% of suitable habitat patches at best. Consider the implications of this for the design of nature reserves, in which suitable habitat may be set aside and yet the target species may only be able to occupy a portion of the reserve as dictated by its demographic potential.

2. Extinction is predicted to occur as a threshold response to habitat loss. Patch occupancy declines gradually and linearly until a critical level of habitat is reached, at which point the population abruptly crashes ($p^* = 0$).

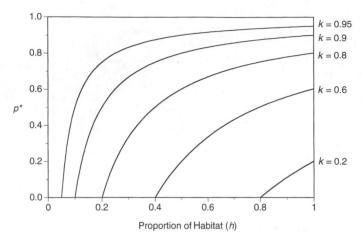

FIGURE 15.4
Extinction thresholds for species with different demographic potentials (k) in random landscapes with different proportions of habitat (h). The demographic potential (k) is a composite variable that incorporates species-specific traits such as net lifetime reproductive output (R_0') and dispersal ability (m). Extinction occurs when equilibrium patch occupancy $p^* = 0$.

3. Species exhibit different thresholds at which they go extinct. Recall that the extinction threshold (h_{crit}) is obtained as $1 - k$. Thus, species are expected to be differentially affected by habitat loss, with some species being very sensitive to habitat loss, such as those with low demographic potentials, owing to a combination of poor dispersal abilities and low reproductive output (e.g., $k = 0.2$).

Conservation Applications

The Northern Spotted Owl (*Strix occidentalis caurina*), federally listed as a threatened species under the Endangered Species Act, inhabits the old-growth coniferous forests (>200 years) of the Pacific Northwest. The conservation saga of the Northern Spotted Owl has been long and controversial (Noon and McKelvey, 1996), stemming from the need to balance timber harvesting with the critical habitat requirements of this species. How much old-growth forest is required for the continued persistence of the Northern Spotted Owl? In other words, what is the extinction threshold for this species? Lande's demographic model was used to predict the extinction threshold for the Northern Spotted Owl (Lande, 1988). At the time, national forests in the regions of western Washington and Oregon contained about 38% old-growth forest ($h = 0.38$). Owls occupied about 44% of the available forested area ($p^* = 0.44$). The demographic potential of this species was thus determined to be $k = 0.79$ (Try it: $k = 1 - [1 - p^*]h$). Subsequently, the extinction threshold for this species is $h_{crit} = 1 - k = 0.21$. Northern Spotted Owls thus are not expected to persist if there is less than 21% old-growth forest on the landscape. Given that implementation of the management guidelines proposed at the time by the USDA Forest Service would have reduced old-growth forest to 7%, the future would have been a bleak one indeed for the Northern Spotted Owl.

Extinction Thresholds on Fractal Landscapes

Lande's model assumes a random distribution of suitable habitat cells (territories) and that individuals (e.g., dispersing juveniles) are able to search for suitable unoccupied territories at random (random distance and direction moved with each dispersal attempt). Lande's demographic model is not spatially explicit and thus cannot take into account how landscape pattern affects extinction thresholds. Given that landscape connectivity is defined as the interaction of organisms with landscape pattern, how would extinction thresholds change as a function of relaxing the underlying habitat distribution (say, from random to a clumped habitat distribution) and search behavior (from random search to constraining individuals to search neighboring cells)? The second half of this lab will permit you to interpret graphical output from a general, spatially explicit modeling approach that combines a metapopulation model with neutral landscape models to explore how habitat loss, apart from the pattern of loss (i.e., fragmentation), affects extinction thresholds for species with different demographic potentials.

The second half of this lab is designed as a graphical analysis and thought exercise to facilitate discussion. The graphs provided for this analysis (Figures 15.5–15.8) were generated within a spatially explicit modeling approach that coupled fractal neutral landscapes with a metapopulation model (With and King, 1999a). Briefly, the objective was to incorporate a fractal distribution of habitat within the context of Lande's (1987) demographic model (Figure 15.4) to explore how extinction thresholds shifted for species with different demographic potentials as a consequence of habitat fragmentation. This involved simulating dispersal as a nearest-neighbor process (individuals moved through adjacent cells of the landscape in search of suitable unoccupied territories) on fractal landscapes, from which dispersal success was calculated as the fraction of individuals ($n = 1000$) that found a territory in $m \leq 50$ dispersal attempts. A mathematical function was then derived to fit these dispersal success curves, which is a function of prior patch occupancy (p) and the amount (h) and clumping (H) of habitat, and which could be substituted for the dispersal success term in Lande's analytical model to solve algebraically for p^*, the proportion of patches occupied at demographic equilibrium (Table 15.3). This resulted in a family of curves that could be compared directly to Lande's results for random dispersal on random landscapes (e.g., Figure 15.5).

EXERCISES

In this exercise, you will interpret graphical output from the model to explore how species with different demographic potentials are affected by habitat loss (h) and fragmentation (H), and under what conditions habitat loss precipitates extinction of the population. Because of the added complexity of quan-

FIGURE 15.5

Extinction thresholds for species on random and fractal landscapes. Populations are near replacement ($R_0' = 1.01$), but species differ in dispersal ability ($m = 2$, 5, or 10). Note that the scaling of the x-axis differs between the top panel ($m = 2$) and the other two panels. The combination of reproductive output (R_0') and dispersal ability (m) gives rise to three species that differ in demographic potential (k). "Random" curves are generated from Lande's (1987) demographic model, which assumes a random distribution of habitat and a random encounter with suitable habitat. A larger, printable version is on the CD.

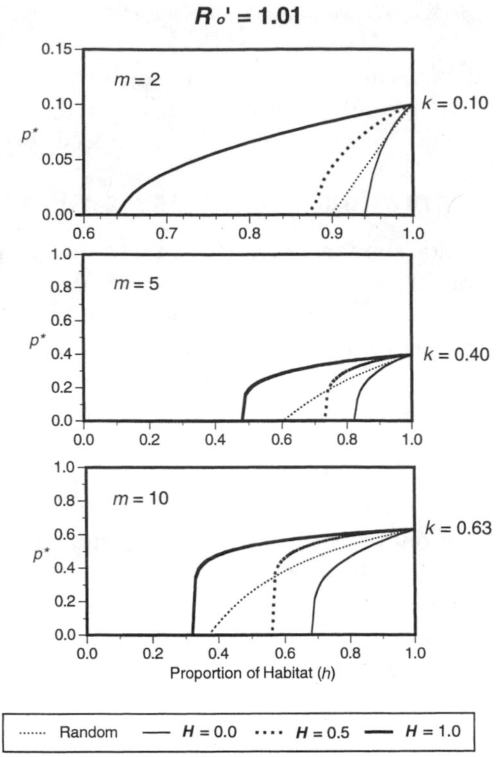

FIGURE 15.6

Extinction thresholds for species with reproductive output $R_0' = 1.10$, but different dispersal abilities (m), on random and fractal landscapes.

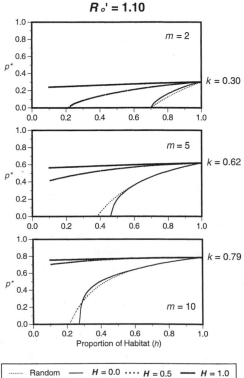

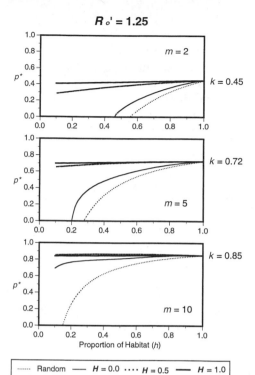

FIGURE 15.7
Extinction thresholds for species with reproductive output $R_0' = 1.25$, but different dispersal abilities (m), on random and fractal landscapes.

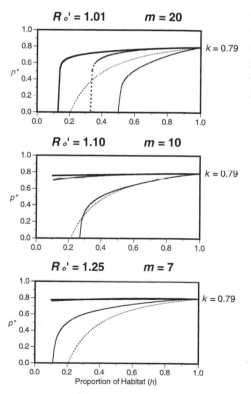

FIGURE 15.8
Comparison of extinction thresholds for species with the same demographic potential (k), but with different combinations of reproductive output (R_0') and dispersal abilities (m).

tifying individual dispersal success on fractal landscape patterns, however, it is no longer possible to treat the demographic potential (k) as a composite parameter as in Lande's model. Instead, it is necessary to tease apart the relative effects of reproductive output (R_0') and dispersal ability (m) in order to make predictions about patch occupancy in fragmented landscapes. In other words, different combinations of R_0' and m can give rise to the same demographic potential (k), but may have very different consequences for the persistence of the metapopulation in a fragmented landscape.

EXERCISE 2.1

Begin by exploring how a species with low reproductive output (R_0') is affected by habitat loss (h) and fragmentation (H). Figure 15.5 displays the expected patch occupancy (p^*, the proportion of suitable habitat patches occupied at demographic equilibrium) for three hypothetical species whose reproductive output is near replacement ($R_0' = 1.01$), but which differ in dispersal ability (poor disperser, $m = 2$; intermediate dispersal ability, $m = 5$, and good disperser, $m = 10$).

Question 2.1. What is the demographic potential (k) of a poor disperser ($m = 2$) with this level of reproductive output? What does this mean?

Question 2.2. What is the extinction threshold (h_{crit}) for this species in a clumped ($H = 1.0$) fractal landscape as opposed to a highly fragmented ($H = 0.0$) one? Remember that the extinction threshold is defined as the level of habitat (h) at which patch occupancy $p^* = 0$. If necessary, design a table to help answer this question.

Question 2.3. How do extinction thresholds for this species in fractal landscapes compare to that predicted by Lande's model of random dispersal in a random landscape? (The "Random" line in Figures 15.5–15.8 is generated by Lande's demographic model.) If necessary, design a table to help answer this question.

EXERCISE 2.2

Still looking at Figure 15.5, examine how increasing dispersal ability (m, the number of sites that can be visited in the search for a suitable unoccupied territory) affects the extinction threshold.

Question 2.4. How does increasing dispersal (m) affect demographic potential (k)? Compare the extinction thresholds (h_{crit}) for a given landscape type (e.g., clumped fractal, $H = 1.0$) among species with different dispersal abilities (m). Are populations of good dispersers expected to persist longer, or go extinct sooner, than populations of poor dispersers under a scenario of habitat loss (decline in h)?

Question 2.5. Does an increase in dispersal ability have a greater effect on population persistence in clumped ($H = 1.0$) or fragmented ($H = 0.0$) landscapes? Compare the magnitude of the shift in the extinction threshold value (h_{crit}) as a function of dispersal ability (m) in different landscapes (clumped fractal vs. fragmented fractal).

EXERCISE 2.3

Now examine a different group of "species" with a higher reproductive output ($R_0' = 1.10$ or $R_0' = 1.25$, Figures 15.6 and 15.7, respectively). Note that each group (defined by a particular R_0') consists of three "species" that differ in dispersal ability.

Question 2.6. How does changing R_0' affect the demographic potential (k)? Compare the value of k across Figures 15.5–15.7 for species with the same dispersal ability (m), but which differ in reproductive output (R_0').

Question 2.7. As reproductive output is increased, what happens to the extinction threshold (h_{crit}) for species in clumped fractal landscapes ($H = 1.0$)? In fragmented fractal landscapes ($H = 0.0$)?

Question 2.8. How do the extinction thresholds (h_{crit}) in fractal landscapes compare to those in random landscapes? Under what scenarios is population persistence greater in all fractal landscapes relative to random landscapes?

EXERCISE 2.4

Finally, let's explore how different combinations of R_0' and m that give rise to the same demographic potential (k) can have different implications for population persistence. Figure 15.8 displays three hypothetical species, each of which has a demographic potential $k = 0.79$, but which differ in their reproductive output and dispersal ability.

Question 2.9. Under what landscape scenarios is a species with this particular demographic potential predicted to go extinct? Is it the same for all species, regardless of R_0' and m? Explain.

Question 2.10. What does this say about our ability to predict a species' response to habitat loss and fragmentation from an estimate of just its demographic potential, k (as was done for the Northern Spotted Owl)?

DISCUSSION QUESTIONS

Answer these questions on additional sheets of paper or discuss as a group.

1. How does habitat fragmentation (H), apart from habitat loss (h), affect the extinction threshold (h_{crit})? Which has the greater effect on population persistence?

2. Fractal landscape patterns are more characteristic of natural habitat distributions than random patterns. Generally speaking, how does population persistence in clumped fractal landscapes ($H = 1.0$) compare with those in random landscapes? Can you offer an explanation for this difference? (*HINT*: Consider how dispersal success might differ in these two landscapes.)
3. What types of species (in terms of R_0' and m) appear to be more sensitive to habitat loss than previously predicted by Lande's demographic model? Can you think of some real-life examples of such species?
4. Which factor has the greatest effect on reducing extinction risk for populations on fractal landscapes: reproductive output (R_0') or dispersal (m)?
5. In light of your response to Discussion Question 4, what would you advise in terms of the conservation of populations on fractal landscapes: the maintenance of high-quality habitat to enhance reproductive success or the creation of corridors to facilitate dispersal success?

ACKNOWLEDGMENTS

My research on the effects of landscape connectivity and the integration of neutral landscape models and metapopulation theory has been supported by the National Science Foundation through grants DEB-9532079 and DEB-9610159. I thank Tony King for his collaboration in the development of the model that generated the output presented in Figures 15.5–15.8.

BIBLIOGRAPHY

Note. An asterisk preceding the entry indicates that it is a suggested reading.

ANDRÉN, H. 1994. Effects of habitat fragmentation on birds and mammals in landscapes with different proportions of suitable habitat: A review. *Oikos* 71:355–366.

DALE, V. H., S. M. PEARSON, H. L. OFERMAN, AND R. V. O'NEILL. 1994. Relating patterns of land-use change to faunal biodiversity in the central Amazon. *Conservation Biology* 8:1027–1036.

FAHRIG, L. 1997. Relative effects of habitat loss and fragmentation on population extinction. *Journal of Wildlife Management* 61:603–610.

GARDNER, R. H., B. T. MILNE, M. G. TURNER, AND R. V. O'NEILL. 1987. Neutral models for the analysis of broad-scale landscape pattern. *Landscape Ecology* 1:19–28.

GARDNER, R. H., AND R. V. O'NEILL. 1991. Pattern, process and predictability: The use of neutral models for landscape analysis. In M. G. Turner and R. H. Gardner, eds. *Quantitative Methods in Landscape Ecology.* Springer-Verlag, New York, pp. 289–307.

GUSTAFSON, E. J., AND R. H. GARDNER. 1996. The effect of landscape heterogeneity on the probability of patch colonization. *Ecology* 77:94–107.

HANSSON, L. 1991. Dispersal and connectivity in metapopulations. *Biological Journal of the Linnean Society* 42:89–103.

LANDE, R. 1987. Extinction thresholds in demographic models of territorial populations. *American Naturalist* 130:624–635.

LANDE, R. 1988. Demographic models of the northern spotted owl (*Strix occidentalis caurina*). *Oecologia* 75:601–607.

LEVINS, R. 1969. Some demographic and genetic consequences of environmental heterogeneity for biological control. *Bulletin of the Entomological Society of America* 15:237–240.

NOON, B. R., AND K. S. MCKELVEY. 1996. Management of the spotted owl: A case history in conservation biology. *Annual Review of Ecology and Systematics* 27:135–162.

PEARSON, S. M., M. G. TURNER, R. H. GARDNER, AND R. V. O'NEILL. 1996. An organism-based perspective of habitat fragmentation. In R. C. Szaro and D. W. Johnston, eds. *Biodiversity in Managed Landscapes: Theory and Practice.* Oxford University Press, Oxford, UK, pp. 77–95.

TAYLOR, P. D., L. FAHRIG, K. HENEIN, AND G. MERRIAM. 1993. Connectivity is a vital element of landscape structure. *Oikos* 68:571–573.

*WIENS, J. A. 1997a. Metapopulation dynamics and landscape ecology. In I. Hanski and M. E. Gilpin, eds. *Metapopulation Biology: Ecology, Genetics and Evolution.* Academic Press, San Diego, pp. 43–62. Explores how landscape structure may often be an important component of metapopulation dynamics and proposes ways in which landscape ecology and metapopulation theory can contribute to an "exciting scientific synthesis."

WIENS, J. A. 1997b. The emerging role of patchiness in conservation biology. In S. T. A. Pickett, R. S. Ostfeld, M. Shachak, and G. E. Likens, eds. *The Ecological Basis of Conservation: Heterogeneity, Ecosystems, and Biodiversity.* Chapman and Hall, New York, pp. 93–107.

WILCOX, B. A., AND D. D. MURPHY. 1985. Conservation strategy: The effects of fragmentation on extinction. *American Naturalist* 125:879–887.

*WITH, K. A. 1997. The application of neutral landscape models in conservation biology. *Conservation Biology* 11:1069–1080. An overview of neutral landscape models and how they might contribute to a variety of applications in conservation biology.

*WITH, K. A., AND T. O. CRIST. 1995. Critical thresholds in species' responses to landscape structure. *Ecology* 76:2446–2459. Numerical analysis coupled with empirical research to explore how individual movement responses to landscape structure affect the resulting patterns of distribution in different species. Recipient of the Award for Best Paper published in the discipline of landscape ecology by the United States Chapter of the International Association for Landscape Ecology in 1996.

WITH, K. A., AND A. W. KING. 1997. The use and misuse of neutral landscape models in ecology. *Oikos* 79:219–229.

*WITH, K. A., AND A. W. KING. 1999a. Extinction thresholds for species in fractal landscapes. *Conservation Biology* 13:314–326. Assumptions inherent in Lande's (1987) original demographic model were relaxed to include limited dispersal on spatially structured landscapes. This synthesis between neutral landscape models and metapopulation theory is the basis for the second part of this lab.

WITH, K. A., AND A. W. KING. 1999b. Dispersal success on fractal landscapes: A consequence of lacunarity thresholds. *Landscape Ecology* 14:73–82.

WITH, K. A., R. H. GARDNER, AND M. G. TURNER. 1997. Landscape connectivity and population distributions in heterogeneous environments. *Oikos* 78:151–169.

CHAPTER 16

INDIVIDUAL-BASED MODELING

THE BACHMAN'S SPARROW

JOHN B. DUNNING JR., DAVID J. STEWART, AND JIANGUO LIU

OBJECTIVES

The long-term survival of a large population is related to the amount of suitable habitat within a diverse landscape. In contrast, the survival of small (e.g., threatened and endangered) populations in restricted landscapes can be dominated by stochastic events and the spatial arrangement of suitable habitat within those landscapes (Hanski and Gilpin, 1991; Hanski, 1998). Population survival can be even more tenuous if the species' suitable habitat is transient in nature. Thus, in a dynamic landscape, population persistence may depend on both the *current* and *future* availability of sufficient suitable habitat. The ability of a species to disperse to new suitable habitat becomes important when the landscape consists of a constantly shifting mosaic of habitat patches of varying suitability (Pulliam et al., 1992). These types of situations are particularly suited to spatially explicit, individual-based population models (Dunning et al., 1995a). In these exercises the student will

1. observe the challenges in managing the natural stochastic variation in size displayed by a small population of a "fugitive species" in a heterogeneous, dynamic landscape;
2. gain an appreciation of the constraints and conflicting priorities often placed on managers;
3. begin to understand the decisions that must be considered when designing a spatially explicit or inexplicit model;

4. understand the utility of individual-based modeling; and
5. determine which landscape features contribute to the long-term persistence of Bachman's Sparrow populations.

Students will use the BACHMAP (Bachman's Sparrow Mobile Animal Population) model, a spatially explicit, individual-based landscape model designed to simulate the consequences of forest management on a population of Bachman's Sparrow (*Aimophila aestivalis*) (Pulliam et al., 1992; Liu, 1993). The model allows the user to populate landscapes with several breeding pairs of sparrows and track the population's annual survival through time. Students will examine the consequences of several model manipulations, including the analysis of population survival on hypothetical maps, as well as on maps of actual landscapes. In addition to the files for this lab on the CD, students will also need printed versions of the data collection tables (which can be printed using Adobe Acrobat Reader).

INTRODUCTION

Many human modifications of landscapes are broad in both spatial and temporal scale and can profoundly affect the organisms occupying these changing landscapes. This is particularly critical when considering **fugitive species,** or those species that depend on ephemeral, constantly changing habitat types. For instance, when a timber management plan is adopted, the distribution of young and old forest will change as forest stands are cut and replanted in new trees. Thus, for organisms that require only a certain age class of forest, the distribution of suitable habitat will change throughout the region during the years in which the management plan is in place. These changing age-class distributions can affect how organisms disperse, select territories, and interact with other individuals as they search for areas of the appropriate age class (Pulliam et al., 1992, Liu et al., 1995). Increasingly, managers are required to determine the impacts of these broad-scale changes on fugitive species (as well as other wildlife and rare and threatened species), ecosystem functions, and other attributes of the environment that society values.

Unfortunately, land managers rarely have the ability to test the long-term and broad-scale impacts of their management plans. While the general effects of a land management strategy may be understood, it is often more difficult to determine the impacts of specific land-use changes on a specific population in a real-world landscape. Will an old-growth specialist such as the Spotted Owl (*Strix occidentalis*) decline if timber cutting is increased by 5% in a certain Forest Service district? If a prairie plant is a poor disperser, can a program of prescribed burning be designed so that plots of suitable ages are always near populations likely to produce seed in a given year? If a pine-forest-endemic bird such as the Bachman's Sparrow can use both old-growth forest stands and newly produced clearcuts, is there a timber harvest strategy that maximizes the amount of suitable habitat for the bird while at the same time generating sufficient income? For each of these questions, a definitive experiment to generate answers quickly in the real world cannot be conducted. In such cases, what is needed is an alternative to fully replicated field experiments (Hargrove and Pickering, 1992).

One such alternative is a **spatially explicit population model (SEPM)**. Models are spatially explicit when the locations of objects (organisms, habitat patches, barriers to dispersal, etc.) are explicitly built into the operation of the model (Dunning et al., 1995a). For example, rather than just knowing that 20% of the landscape is in forest, the exact spatial locations of the forested areas are known and incorporated into the model, often by constructing a map of landscape features. Such landscape maps are often designed as a grid of square cells (or hexagons), and habitat attributes (habitat type, age, suitability) can be assigned to each cell. Real-world landscapes can be simulated by creating a grid of cells that roughly capture the heterogeneity of habitat patches found in a real landscape (Liu et al., 1995; see also Chapter 9, Neutral Landscape Models). The landscape map can be made dynamic, changing predictably (or unpredictably) according to rules built into the model.

Spatially explicit models differ from **spatially inexplicit models** in a variety of ways (Dunning et al., 1995a). The most important is the way landscape features are used in the model. Many nonspatial landscape models include landscape metrics in their operation, but the locations of landscape features are not used in the model. For instance, a model might judge suitability of a landscape for a species by tracking the amount of suitable habitat present. This is a measure of landscape composition (not spatial configuration—see Chapter 7, Understanding Landscape Metrics I), which has been found to be important in many landscape analyses (Dunning et al., 1992; Rosenberg et al., 1999). This kind of model is a landscape-scale model, but it is not spatially explicit because the locations of the habitat patches are not measured or important to the model. A spatially explicit model might define suitability for a species based on the placement of suitable habitat within 10 kilometers of a specific location in the landscape (e.g., a breeding site or foraging location).

Spatially explicit models can require enormous amounts of data, which are rarely available in most situations. Therefore, if one can conduct a landscape analysis without building a spatially explicit model, it is usually better to avoid the complexity of a fully spatial model (Wennergren et al., 1995). For instance, analyses of the landscape features that appear to be important to a variety of nongame birds in the western United States revealed that many species were responsive to the composition of the landscape (i.e., how much of an area was covered with specific types of forest), but that the distribution of the habitat patches was less important (McGarigal and McComb, 1995). For such species, landscape composition models can accurately track the effects of landscape change—a fully explicit model specifying the exact location of each patch is unnecessary. For many organisms with moderate to good dispersal ability, it appears that landscapes composed of 25% or more suitable habitat rarely show strong landscape effects on population dynamics (Fahrig, 1997). In these situations, a fully explicit landscape model is likely unnecessary.

Many SEPMs are **individual-based;** that is, the model follows individual organisms as they move across a heterogeneous landscape. Individuals gain habitat-specific fecundity or mortality traits based on where they are on the landscape. Individual-based spatial models parallel (and can borrow from) the strong tradition and well-developed literature of individual-based *nonspatial* demographic models (Pulliam and Dunning, 1997). Many SEPMs have been created for threatened and endangered species when tracking each individual in the landscape may be a management concern, and identifying the impacts of landscape change on specific age classes or cohorts is important. In fact, it is often true that the extensive demographic and dispersal data needed for these models are only available for rare and endangered species. Funding is often unavailable for gathering basic natural history data for more common species.

SEPMs can also be population based. Instead of tracking individuals across the landscape and simulating mortality and fecundity of each organism, one can assume that each patch in the landscape supports a population. Population-level parameters such as birth rates and death rates can be modeled, simulating both density-dependent and density-independent factors when appropriate. Instead of monitoring the movements of individuals, the program can calculate immigration and emigration rates. This type of SEPM is more appropriate for species with high population densities and high fecundity and mortality rates, when tracking each individual is impractical (both in the simulation and in the real world). This type of SEPM can be especially important in building a metapopulation model for relatively common organisms.

We should emphasize that spatially explicit models cannot be used to make direct experimental tests of the management questions posed at the start of this chapter. A direct test of whether Spotted Owls will increase or decrease when a specific management strategy is implemented is logistically impossible—one would have to implement the strategy in multiple, identical, replicate landscapes (which could not be found in the real world) and compare the owl population dynamics to those in control landscapes. Computer models provide an investigative tool for exploring *possible* consequences of management changes, given that direct tests are usually not possible.

Bachman's Sparrow

Bachman's Sparrow, a resident species of the southeastern United States, is a species of management interest because it has declined in range and abundance for much of the 20th century (Dunning, 1993). The sparrow also presents a management challenge as a fugitive species in a rapidly changing, fragmented landscape. Bachman's Sparrow appears to be most attracted to pine stands with a dense layer of grasses and forbs on the ground and relatively few tall shrubs (Dunning and Watts, 1990).

Historically, Bachman's Sparrows were probably most common in old-growth (>100-year) pine forests (especially longleaf pine, *Pinus palustris*, on the southeastern coastal plain) where a wide spacing of canopy trees and frequent fires support a dense community of ground-layer plants. Over 90% of these old-growth forests have been harvested, however. In their place, timberlands often support younger (20- to 40-year-old) pine plantations. These denser forests often have a dense layer of pine needles on the ground with few grasses and forbs.

Increasingly, suitable conditions within these forest districts are found only in very young pine stands, especially newly planted clearcuts (Dunning and Watts, 1990; Dunning et al., 1995a). For the first few years after planting, such areas have a dense layer of grasses and forbs and few shrubs until the planted pines grow large enough to shade out the grasses. In the coastal plain of South Carolina and Georgia, pines planted in new clearcuts may grow so fast that stands may be suitable for only 3 to 4 years postplanting. Thus, instead of occupying a landscape of old forest where most areas are burned frequently and suitable in most years, the sparrows are forced into a fugitive existence, moving around frequently to find new patches of short-lived clearcut habitat. This landscape change may be in part responsible for the species' decline, although this hypothesis is difficult to prove (Dunning and Watts, 1990).

The BACHMAP Model

BACHMAP is a spatially explicit, individual-based model designed to explore the consequences of forest management on a population of Bachman's Sparrows (Pulliam et al., 1992; Liu et al., 1995). BACHMAP simulates year-by-year changes in a population of Bachman's Sparrows that occupies certain age classes of pine forest. Although this sparrow is often found in managed pine plantations, it breeds successfully only in a narrow range of habitat types—mature forest stands and recently planted clearcuts. The displayed map is color-coded to indicate the relative suitability of each cell (or pixel) for sparrow breeding. A cell is 2.5 hectares in size, representing the territory of a breeding pair. In the simulation, differences in breeding success in different areas are treated as probabilities that individual breeding pairs will produce offspring. Additional colors indicate the presence of a breeding pair within a habitat unit and the paths of dispersing sparrows. At the close of the simulation, the program displays a chart showing the change in population size through time.

While agencies such as the U.S. Forest Service consider the bird to be a management priority, maintaining species of management concern is not the only priority of the Forest Service. District rangers are also under pressure to ensure that timber is harvested off their forest districts in an economically justifiable manner. Therefore, later versions of BACHMAP also track the income generated from timber harvest from the forest landscape. This economic function was originally developed in a related model called ECOLECON (Liu et al., 1994; see **For Further Study** at the end of this chapter for details). Thus, the model can be used to help managers balance two sometimes conflicting priorities—meeting economic and ecological goals.

MODEL INPUT

The model has a variety of input parameters that are specific to the biology of the sparrows, as well as other parameters that can be adjusted to model certain scenarios. The most fundamental parameters to understand are as follows.

Base Maps. Preexisting maps can be used as input to the model, or maps can be created using the model. Two sample maps, representing the same landscape at different times, are included in this demonstration package: (1) Savannah River landscape, 1989, and (2) Tornado Alley, 1991. The boundaries of this region closely follow the compartment boundaries actually used at the Savannah River Site, a Department of Energy nuclear facility near Aiken, South Carolina. In the maps of these real-world landscapes (see Exercise 3), north is to the top of the screen. ArcInfo routines to create new maps based on GIS coverages of other landscapes are available from the second author. Maps created by the model are always rectangular in shape. All maps consist of two files describing the landscape topology (**generate.gis**) and the sparrow population distribution (**generate.adl**).

Habitat Suitability. Each cell is ranked according to its habitat suitability, which reflects the probability that a breeding pair will produce offspring in a given age class of forest. The range of forest ages considered "highly suitable" or "less suitable" can be altered in the model.

Bachman's Sparrow Dispersal Characteristics. Several attributes of adult and juvenile dispersal are included in the model. The site fidelity of adults is variable; that is, users can change whether adult birds tend to return to the same site for breeding. For juveniles, important considerations are whether the juvenile can inherit its parents' territory or whether the juvenile must disperse to find appropriate habitat. Other factors built into the model include the juvenile's ability to "see" suitable habitat from a distance and "judge" its suitability. These variables affect whether dispersing individuals move toward the closest suitable habitat. Also included are probabilities of changing flight paths, dispersal mortality, and whether an individual may land on a habitat cell already occupied by a breeding pair.

Management Strategy. Several options exist for defining the harvest management strategy, such as whether to continue harvesting after sparrow extinction. There is a regular harvest option, and the length (in years) of each rotation can be specified.

MODEL OUTPUT

Chart of Population Survival. At the close of each simulation, a data file is constructed that contains the sparrow's annual population sizes. These numbers are plotted as a chart in which the horizontal axis is the simulation year, and the left-hand vertical axis is the number of breeding pairs at the close of the sparrows' annual cycle. If more than one replicate data set has been collected, the population track from each replicate will be displayed. The raw population data are found in a file at

C:\MODELS\BACHMAP\TEMP\METAPOP.REP

Economic Data. A separate output data file contains the annual economic return based on the simulated harvest of trees. The values of economic performance for each run of the model are plotted on the population survival chart, scaled to match the right-hand vertical axis. The path to the file containing the raw economic data is

C:\MODELS\BACHMAP\TEMP\ECONOMY.REP

Instructions for BACHMAP

Installation. Before running BACHMAP, it must be first installed on you computer. Double-click on the file **mapinst.exe** to unzip these onto your computer to a default directory of C:\models\bachmap.

Running BACHMAP. In Windows, double-click the Bachmap icon under C:\models\bachmap. In DOS, use the CD command (change directory command) to change to the BACHMAP directory, and type "BACHMAP."

System Requirements. BACHMAP simulations use graphics routines that may be incompatible with running Microsoft Windows. If you experience problems, run the program in DOS mode only, or a fully expanded DOS window. Versions for Macintosh and Unix platforms are available upon request from the second author. The BACHMAP menus require an environment size of at least 1024 Kb. If you experience trouble displaying the menus, try inserting this line in your CONFIG.SYS file:

SHELL=C:\COMMAND.COM /E:1024 /P

In Windows 95 or Windows 98, verify that the initial environment for the file BACHMAP.BAT is set to 1024 or above by right-clicking on the file icon, selecting **Properties** from the pull-down list, opening the **Memory** tab, and examining the **"Initial environment"** setting. It should *not* read "auto". Contact the authors if you experience problems, particularly in a Windows 2000 environment.

Main Menu. After starting BACHMAP, you should see a main menu of choices as shown below. At the top of the screen will be printed the current landscape loaded in memory, if any.

```
BACHMAP—Bachman's Sparrow Mobile Animal Population Simula-
tion
  GIS files loaded from the directory
    0—Quit
    1—Introduction
```

```
   2—Create a new landscape
   3—Select a landscape
   4—Edit model parameters
   5—Run the simulation
   6—Display population trends
   7—Display a landscape
Choose a number:
```

Submenus. After choosing a number from 0 to 7 from the main menu, several more options are displayed. Be sure to familiarize yourself with the different options (and briefly note the suboptions!) as you read through this section. The default answers are shown in parentheses. Press Enter after each question to select the default value, or type a new value.

Option 0 in the main menu is self-explanatory. Option 1 provides a brief introduction to the model. In the following sections, we discuss the remaining options in more detail.

Option 2: Create a New Landscape

This option allows the user to create a hypothetical map to be used as the base map for analyses and consists of the following:

```
To Build the Landscape, Please Answer the Following Ques-
tions:
Accept default artificial landscape structure: Y/N?
```

SUBOPTIONS IF ANSWERING "NO":

```
quadrant # within metalandscape:
number of rows for entire landscape:
number of columns for entire landscape:
number of cells in each stand:
number of cell rows in each stand:
randomly distribute stands:
number of mature stands:
  number of cells in each mature stand:
  number of cell rows in each mature stand:
  randomly distribute mature stands:
number of non-pine stands:
probability that mature stands are occupied:
probability that 1-2 year stands are occupied:
probability that 3-5 year stands are occupied:
Number of stand age classes =
Are age classes exactly even:
Are excess age groups treated as hardwoods:
seed for random number generator =
Are you satisfied with these answers: Y/N?
```

Please note that entering a zero for the random number generator guarantees a different landscape each time. Each nonzero seed produces a unique landscape. The "number of cell rows" suboption indicates how the cells in a given stand are organized into a block.

Option 3: Select a Landscape

This option allows the user to select one of several landscapes stored in the program:

```
0—return to main menu
1—Savannah River landscape, 1989
2—Experimental landscape with a powerline right-of-way
3—Tornado Alley, 1991
4—user defined
```

Option 4: Edit Model Parameters

This option allows the user to change a variety of model parameters dealing with the characteristics of the landscape, the sparrow, or the management regime. The user can also change the performance of the model itself (e.g., the number of replications that will be done of a particular model run). You will be given the following prompts:

```
0—return to main menu
1—Edit the current parameter file
2—Load the pre-set parameter file
```

If the user wishes to change the default parameter values then select "I" (edit the current parameter file) and then select "O" (overwrite). The default value for each parameter value is shown in parentheses. Press Enter after each question to select the default value, or type a new value:

```
Simulation run time (in years) =
Number of replicates =
Accept default values for habitat suitability: Y/N?
Accept default dispersal characteristics: Y/N?
Accept default management strategy: Y/N?
Accept default graphics display options: Y/N?
seed for random number generator = (0)
Are you satisfied with these answers: Y/N?
```

If the user selects "N" for any of the questions that begin "Accept default . . . ," a new list of suboptions will be presented, allowing the user to change specific parameters. Again, please note the consequences of choosing a nonzero value for the random number seed. Each nonzero seed produces a unique—but repeatable—stream of random numbers, so that all simulations on a given landscape with the same model parameters and sparrow population, starting with the same nonzero random number seed, will repeat exactly.

Option 5: Run the Simulation

(a) Wait for the title screen to appear and for the landscape to load. The screen display is a bird's-eye view of a landscape, composed of a packed array of hexagons. Each hexagon (cell) is color-coded to indicate the habitat suitability of the cell and whether the cell is occupied by a sparrow. On the screen the colors are defined in a legend that appears at the upper right corner. Cells dominated by unsuitable habitats are black (unsuitable pine age classes) or blue (bottomland hardwoods). A dispersing bird may briefly occupy unsuitable habitat as it moves through the landscape, but it cannot "decide" to settle there. However, a territorial bird that survives the nonbreeding season may choose to remain in its territory from the previous year. If its territory was marginal habitat in the past year, then the bird would be now found in unsuitable habitat.

Only one sparrow (pair) can occupy a cell at any given point in time. Occupied cells are color-coded green (mature), yellow (suitable clearcuts), red (marginal clear-cuts), and purple (poor). Unoccupied cells dominated by pine forest are colored white (mature pine forest—the most suitable age class) or various shades of gray (the less suitable clearcut age classes). Although the model simulates the movements and demography of female sparrows only, occupancy by a sparrow is assumed to indicate occupancy by a male–female pair. Thus, an individual settling in a cell of high-quality habitat is assumed to gain the reproductive success associated with that habitat, without specifically modeling pairing success.

(b) Press Enter to begin the simulation.

(c) *Step-by-step mode.* Initially, the program will show you every individual that must change status (by dying, inheriting a cell, or dispersing). Press Enter or the space bar to move through the landscape cell by cell, showing each individual in turn. The individual being considered at any given point is marked on the map with a flashing colored cell, and the characteristics of that cell and the individual are found at the bottom of the legend along the right margin.

(d) *Continuous mode.* To eliminate the step-by-step process, hit the "+" key at any time. When in continuous mode, you can hit the "+" key or the "−" key to slow down or speed up the movements. Watch the legend where it says "Draw Delay" to see how this works. On some computer systems, this option has proven to be balky. If the simulation seems to be running too fast, hit the "−" key once or twice to slow the movements, then wait for the next annual cycle to see if the performance improves.

(e) When the simulation is done, press any key to return to the BACHMAP menu.

Option 6: Display Population Trends

This produces a chart of annual survival of the sparrow population.

Option 7: Display a Landscape

This options allows the user to view the landscape currently in use including the pre- and post-simulation landscapes.

EXERCISES

You will run the BACHMAP simulation using default values for most of the simulation parameters, except where indicated in each exercise. By entering a series of parameters, you will create small, hypothetical heterogeneous landscapes for use in Exercises 1 and 2. In the last exercise an actual landscape is used as the basis for a simulation.

WARNING: The model does not revert to default conditions after each simulation. If you change a parameter value for a simulation, you should record the original (default) parameter values along with your changes prior to running the simulation. You should then restore the parameter to its original value if the next simulation requires default conditions.

After several simulations with the graphics operational (i.e., so that the landscape map appears and all changes are displayed with each simulation), users may wish to run the simulations without viewing the graphics. This option can be found under the "Use default graphics display" suboption under the model parameters menu mentioned later. Turning off the graphics will speed up the simulations; however, we suggest that users run at least some simulations with the graphics on to make sure the simulations are doing what the users expect. In one set of simulations, we discovered that all dispersers were moving only in a SW direction, instead of randomly as we thought. The mistake was the result of a typing error in a dispersal variable and would have been extremely hard to discover if the graphics were not available as an accuracy check.

EXERCISE 1
Bachman's Sparrow Survival on a Hypothetical Landscape: The Basics

The first BACHMAP simulation illustrates the survival of Bachman's Sparrow within a small, hypothetical landscape. Only the default model parameters will be used to generate the landscape, choose the dispersal characteristics, and set the harvest schedule. The landscape will be a 400-cell grid with 100 stands of four cells each. Six of those stands will be "old-growth" forest (100 years old); the rest will be randomly chosen ages between 0 and 20 years. The harvest schedule is a fixed, 20-year rotation: all stands that reach 20 years old are automatically harvested. All older stands (including the mature stands) are left alone.

1. At the main menu at the top of the screen will be printed the current landscape loaded in memory, if any. For this exercise your next step will be to create a new landscape.

2. Select option 2, **"Create a new landscape."**
   (a) The default value is shown in parentheses. Press Enter after each question to select the default value, or type a new value:

```
    Accept default artificial landscape structure: Y/N?
     (Yes)
    Number of stand age classes = (20)
    Are age classes exactly even: Y/N? (Yes)
    Are excess age groups treated as hardwoods: Y/N?
     (No)
```

(b) Select a nonzero random number seed:

```
    seed for random number generator = (0) 63084
```

(a) Press Enter to return to the BACHMAP menu:

```
    Are you satisfied with these answers: Y/N? (Yes)
```

Since you are creating a new landscape, you do not need to use option 3, "Select a landscape," at this point.

3. Select option 4, "**Edit model parameters.**"
   (a) If the parameter file exists, select "O" (overwrite).
   (b) The default value is shown in parentheses. Press Enter after each question to select the default value, or type a new value:

```
    Simulation run time (in years)= (20)
    Number of replicates = (1)
    Accept default values for habitat suitability:
     Y/N? (Yes)
    Accept default dispersal characteristics: Y/N?
     (Yes)
    Accept default management strategy: Y/N? (Yes)
    Accept default graphics display options: Y/N?
     (Yes)
    seed for random number generator= (0)
    Are you satisfied with these answers: Y/N? (Yes)
```

4. Select option 5, "**Run the simulation.**"
   (a) Wait for the title screen to appear and for the landscape to load.
   (b) Press Enter to begin the simulation.
   (c) When the simulation is done, press any key to return to the BACHMAP menu.

5. Select option 6, "**Display population trends.**"
   (a) Answer Question 1 at end of *this* exercise.
   (b) Press any key to return to the Bachmap menu.

The second simulation entails changing the amount of mature forest stands present initially on the hypothetical landscape:

1. Use option 2 (Create a new landscape) to change the number of mature forest stands in the hypothetical landscape. Accept the default artificial landscape structure. Note that you should keep all other options (such as harvest rotation) the same so that you are testing just for the effect of mature stands. The default values provided results for an initial landscape containing six old-growth stands.

2. Choose four other values for the initial number of old growth stands (or percent of landscape in old-growth).

3. Plot the population size versus time for each of the five alternative scenarios. Use at least three replicates for each of the scenarios. Record your observations in Table 16.1 from the CD.

Question 1. When the default values of the model were used, how would you describe the simulated 20-year trends in population change?

Question 2. What is the effect of changing the amount of mature forest stands in the landscape on sparrow abundance and variability? Are mature stands necessary for the long-term population survival of Bachman's Sparrow?

EXERCISE 2
Bachman's Sparrow Survival: Creating Complicated Hypothetical Landscapes

In this exercise you will simulate the population dynamics of Bachman's Sparrow on a 2500-hectare hypothetical landscape. The initial landscape contains four 100-year-old pine stands (each 22.5 ha). The remaining landscape consists of 10-hectare pine stands ranging in age from 0 to 39 years. After the simulation begins, all stands initially less than 30 years old are cut after they reach age 29 and immediately replanted. Stands that are initially older than 30 years are allowed to mature. Sparrows are assumed to breed in young stands (1–3 years old) and mature stands (80–110 years old).

As in the first BACHMAP example, you will use the landscape generator to construct a hypothetical landscape and will enter parameter values to create a baseline simulation scenario. After running the simulation using the default values, you will adjust other parameters to answer several specific questions involving dispersal and extinction probabilities. A set of simulation runs will generally be required to answer each question.

1. Option 2—Create a new landscape (do NOT accept the default artificial landscape structure):

```
quadrant # within metalandscape:                    0
number of rows for entire landscape:                40
number of columns for entire landscape:             30
number of cells in each stand:                      4
number of cell rows in each stand:                  2
randomly distribute stands:                         Yes
number of mature stands:                            4
   number of cells in each mature stand:            9
   number of cell rows in each mature stand:        3
   randomly distribute mature stands:               Yes
number of non-pine stands:                          0
probability of occupancy in mature stands:          0.5
probability of occupancy in 1-2 year stands:        0.2
probability of occupancy in 3-5 year stands:        0.2
```

```
Number of stand age classes:              40
are age classes exactly even:            No
seed for random number generator:        46668
```

2. Edit model parameters:

```
Simulation run time (in years):          50
Number of replicates:                     1
Accept default values for habitat suitability:  No
  number of distinct breeding success windows
  (15 max):                               2
  habitat suitability 2:OK 3:good 4:best:  4
  beginning stand age:                    80
  ending stand age:                       109
  habitat suitability 2:OK 3:good 4:best:  3
  beginning stand age:                     1
  ending stand age:                        3
Accept default dispersal characteristics:  No
  Adults have high site fidelity:        No
  Allow juveniles to inherit:            Yes
  Let juveniles anticipate future good habitat:  No
  Differentiate between good cells:      No
  Do boundaries absorb dispersers:       No
  How far (rings) that a juvenile can see=:  1
  probability of path direction change=:  0.1
  Avoid occupied cells:                    1
  Avoid visited cells:                     1
  Avoid hardwood regions:                  1
  Cell-by-cell dispersal mortality=:       0.02
Accept default management strategy:       No
  Continue harvesting if animals go extinct:  Yes
  regular harvest:                        Yes
  length (years) of each rotation:         30
Accept default graphics display options:  Yes
  seed for random number generator:       47318
```

3. Run the simulation and record your observations about the baseline population response to these conditions in Table 16.2 from the CD.

4. Edit the simulation parameters to achieve the following goals. Run the simulation after each change, returning any other modified parameters to their default values. Then, use your results to answer the questions that follow. Enter your results for the next several steps in Table 16.2 in rows a through e.

   (a) Reduce the ending stand age for the second breeding success window to 2 years (instead of 3).

   (b) Add a third breeding success window that is of "OK" quality, for stand ages 3 to 5 years.

   (c) Create a new landscape with zero mature stands.

   (d) Increase the number of replicates to 50.

   (e) Change the rotation length to 40 years, and then 50 years.

Question 3. Using the initial baseline simulation, what will happen to the population after 10 years when the original mature stands become too old to support Bachman's sparrows?

Question 4. How do the sparrow population dynamics change if they can only breed in mature stands plus stands that are (a) 1–2 years old, (b) 1–3 years old, or (c) 1–5 years old?

Question 5. Can the Bachman's sparrow exist as a "fugitive species" in the absence of mature pine stands?

Question 6. What would be the consequence of increasing the rotation length to 40 or 50 years?

Question 7. How variable are the model predictions when you consider the results of multiple simulation runs? How can replicate simulations be used to bracket predictions or compute the likelihood of a particular event (such as extinction)? How might you determine how many replicates should be run?

Question 8. Based on the set of simulations conducted in Exercise 2, where might intervention for enhancing habitat quality or changing harvest strategies be most effective for increasing the likelihood that the sparrows would persist?

EXERCISE 3
Using Maps of Real Landscapes: The Savannah River Site

Principles that are easy to demonstrate in artificial landscapes may not be obvious when considering real landscapes. In addition, the managers of real landscapes often do not have the luxury of conducting controlled experiments with their land resources. Many management techniques are developed by trial and error and may not in fact be the optimal strategy when considering multiple management goals. Thus, exploring the consequences of potential management changes on the actual landscapes on which the management would be implemented is potentially a powerful tool for land managers.

The Savannah River Site

This BACHMAP exercise illustrates sparrow population dynamics in a close approximation of a real landscape, as mapped in ArcInfo, a geographic information system (see Chapter 3, Introduction to GIS for more information). The landscape consists of about 5900 hectares (14,600 acres) of managed pine woodland and deciduous bottomlands found on the Savannah River Site (SRS), a Department of Energy facility in the coastal plain of South Carolina. The U.S. Forest Service manages the majority of the SRS for multiple use, including timber harvest and biodiversity conservation. The map you will be using is derived from compartment maps and stand information supplied by the Forest Service. Opportunities will be presented to modify the planned harvest during the simulation based on observed trends in population size.

1. Select option 3, "**Select a landscape.**"

   Choose landscape 3: Tornado Alley, 1991.

2. Select option 4, "**Edit model parameters.**"
   (a) If the parameter file exists, select "O" (overwrite).
   (b) The default management strategy consists of the default values for management decisions, shown in parentheses following each suboption. Press Enter after each question to select the default values, down to the following questions:
   (c) Accept default management strategy: (Y/N)? (No)

   ```
   Continue simulation if animals go extinct: Y/N? (NO)
   regular harvest: (Y/N) (YES)
   Length (years) of each rotation = (20)
   Discount rate = (0.050000)
   ```

   (d) The remaining questions should be given default values:

   ```
   Accept default graphics display options: Y/N? (Yes)
   ```

3. Run the simulation with these parameter values. Record the population sizes and average annual harvest value for this simulation run and all subsequent simulations in Table 16.3 from the CD.

Question 9. Did your population go extinct with the default settings? What was the average population size at the end? Was the population increasing or decreasing at the end?

Question 10. Change the rotation length and see how that affects the sparrow population. If the population goes extinct, run several replications to see what percentage of the time extinction occurs.

Question 11. Increase the simulation length from 20 years to 50 or 100 years. How does the population respond over the long term?

Question 12. Do you see a relationship between rotation length and average annual harvest value?

Question 13. Under what conditions would you expect a profitable timber harvest to be obtained without endangering the Bachman's sparrow population on this landscape? To answer this question, you will need to find a compromise between the strategy that will maximize income off the property and the strategy that will maximize the sparrow population. Test your ideas by designing a harvest plan with a positive long-term economic yield, and modify the management parameters of the model to implement that plan. Run the simulation for 20 years with ten replicates and report the minimum, mean, and maximum final sparrow population sizes. How do your results compare with those of others in the class?

Question 14. Based on your experience in exploring the behavior of Bachmap, consider the benefits and limitations of developing and implementing an SEPM. Under what circumstances, or for what types of questions, is an SEPM most appropriate? What caveats should be understood when the model results are interpreted? Have these simulations provided output that might direct management or identify important areas for additional field research?

FOR FURTHER STUDY

Solving conflicts between ecological and economic goals requires simultaneous studies on the ecological and economic consequences of management strategies. To search for a balance between species conservation and economic returns, ECOLECON (an ECOLogical–ECONomic model) has been developed based on the original BACHMAP model (Liu, 1993). ECOLECON simulates both sparrow population dynamics and economic cash flows in response to landscape structure and timber harvest in managed forests. The model calculates indices used by foresters to estimate long-term economic returns, such as net present value (NPV). Further details on ECOLECON can be found in Liu (1993) and Liu et al. (1994). The model is available from the third author.

BIBLIOGRAPHY

Note. An asterisk preceding the entry indicates that it is a suggested reading.

*CASAGRANDI, R., AND M. GATTO. 1999. A mesoscale approach to extinction risk in fragmented habitats. *Nature* 400:560–562. Argues that spatially explicit models can be cumbersome and too data intensive. On the other hand, classic macroscale metapopulation models ignore data on the distribution of local abundances. The paper presents an alternative midscale approach for metapopulations subjected to demographic stochasticity, catastrophes, and habitat loss.

DUNNING, J. B. 1993. Bachman's sparrow, *Aimophila aestivalis.* In A. Poole, P. Stettenheim, and F. Gill, eds. *The Birds of North America*, No. 38. The Academy of Natural Sciences, Philadelphia, Pennsylvania; and The American Ornithologists' Union, Washington, DC, pp. 1–23.

DUNNING, J. B., R. BORGELLA, K. CLEMENTS, AND G. K. MEFFE. 1995a. Patch isolation, corridor effects, and avian colonization of habitat patches in a managed pine woodland. *Conservation Biology* 9:542–550.

DUNNING, J. B., B. J. DANIELSON, AND H. R. PULLIAM. 1992. Ecological processes that affect populations in complex landscapes. *Oikos* 65:169–175.

*DUNNING, J. B., D. J. STEWART, B. J. DANIELSON, B. R. NOON, T. L. ROOT, R. H. LAMBERSON, AND E. E. STEVENS. 1995b. Spatially explicit population models: Current forms and future uses. *Ecological Applications* 5:3–11. A summary of the uses of spatially explicit models, this paper is one of several papers on the pros and cons of this modeling technique published together as a Special Feature in this issue of *Ecological Applications.*

DUNNING, J. B., AND B. D. WATTS. 1990. Regional differences in habitat use by Bachman's Sparrows. *Auk* 107:463–472.

FAHRIG, L. 1997. Relative effects of habitat loss and fragmentation on population extinction. *Journal of Wildlife Management* 61:603–610.

HANSKI, I. 1998. Metapopulation dynamics. *Nature* 396:41–49.

HANSKI, I., AND M. GILPIN. 1991. Metapopulation dynamics: Brief history and conceptual domain. *Biological Journal of the Linnean Society* 42:3–16.

HARGROVE, W.W., AND J. PICKERING. 1992. Pseudoreplication: A sine qua non for regional ecology. *Landscape Ecology* 6:251–258.

LIU, J. 1993. ECOLECON: A spatially explicit model for ECOLogical-ECONomics of species conservation in complex forest landscapes. *Ecological Modelling* 70:63–87.

*LIU, J., F. CUBBAGE, AND H. R. PULLIAM 1994. Ecological and economic effects of forest structure and rotation lengths: Simulation studies using ECOLECON. *Ecological Economics* 10:249–265. Reprinted in R. Costanza, C. Perrings, and C. Cleveland, eds. *The International Library of Critical Writings in Economics—The Development of Ecological Economics.* Edward Elgar Publishing Limited, Cheltenham, UK. Uses ECOLECON to address both the impact of changing management scenarios on both economic goals (timber income) and ecological dynamics (sparrow population trends). The paper demonstrates how such factors can be balanced and optimized.

*LIU, J., J. B. DUNNING, AND H. R. PULLIAM. 1995. Potential impacts of a forest management plan on Bachman's sparrow (*Aimophila aestivalis*): Linking a spatially explicit model with GIS. *Conservation Biology* 9:62–79. Uses the BACHMAP model to evaluate the long-term effects of a U.S. Forest Service management plan in a real forest landscape, assessing whether the plan is likely to meet management objectives for species that are not the direct target of management strategy.

McGARIGAL, K., AND W. C. McCOMB. 1995. Relationships between landscape structure and breeding birds in the Oregon Coast Range. *Ecological Monographs* 65:235–260.

PULLIAM, H. R., AND J. B. DUNNING. 1997. Demographic processes: Population dynamics on heterogeneous landscapes. In G. K. Meffe and C. R. Carroll, eds. *Principles of Conservation Biology*, 2nd ed. Sinauer and Associates, Sunderland, Massachusetts, pp. 203–232.

PULLIAM, H. R., J. B. DUNNING, AND J. LIU. 1992. Population dynamics in complex landscapes: A case study. *Ecological Applications* 2:165–177.

ROSENBERG, K. V., J. D. LOWE, AND A. A. DHONDT. 1999. Effects of forest fragmentation on breeding tanagers: A continental perspective. *Conservation Biology* 13:568–583.

*WENNERGREN, U., M. RUCKELSHAUS, AND P. KAREIVA. 1995. The promise and limitations of spatial models in conservation biology. *Oikos* 74:349–356. Several authors have urged caution in developing a heavy reliance on spatially explicit models in conservation biology. This article summarizes many of the concerns that have been raised. For more information on this debate, we suggest readers consult the variety of papers on fragmentation models published by Lenore Fahrig and colleagues.

ECOSYSTEM PROCESSES AT BROAD SCALES

Linking ecosystem and landscape ecology is an emerging challenge in ecology. Here we examine two topics that have been fostered by the integration of ecosystem and landscape ecology. Building on the strong background of organismal response to landscape pattern (see Section V), ecologists have examined organisms as drivers of landscape structure and have found important effects on ecosystem processes. Chapter 17, Feedbacks between Organisms and Ecosystem Processes, examines several systems in which feedbacks between herbivores and landscape structure influence ecosystem processes (such as primary productivity and nutrient cycling) on landscapes. Another area in which both ecosystem and landscape ecologists have been actively engaged is watershed modeling, linking land-use patterns to runoff and water quality in nearby aquatic systems. Chapter 18, Modeling Ecosystem Processes, explores some of the limitations and assumptions inherent in simulating such scenarios, but more important, provides students with a simple simulation environment in which to conceptualize and build their own model and perform their own queries of ecosystem processes operating at broad scales.

CHAPTER 17

FEEDBACKS BETWEEN ORGANISMS AND ECOSYSTEM PROCESSES

LINDA L. WALLACE AND STEVE T. GRAY

OBJECTIVES

The fact that the structure of the environment affects organisms within a system is so well known as to be an axiom of ecology. However, precisely *how* system structure does this is less well understood. Further, organism behavior may have strong feedbacks onto system structure and function. This exercise presents an examination of models and field data to explore interactions between organisms and ecosystem structure at several spatial scales, particularly the landscape scale. This is a critical first step in understanding how any given landscape may function and how perturbations (either human-induced or "natural") may influence the functioning of that landscape and the ecosystems contained within it. The primary objectives in this exercise are to help students

1. understand how ecosystem processes and landscape structure influence organisms at landscape scales, and how organisms influence ecosystem dynamics at equally broad scales;
2. appreciate how the interaction between organisms and their landscapes creates and maintains landscape heterogeneity;
3. understand and discriminate between positive and negative feedbacks; and
4. gain exposure to a diversity of approaches to analyzing the interactions between organisms and ecosystems or landscapes.

In Part 1, you'll examine of model of how fire patterns in Yellowstone National Park influence the winter survival of elk and bison, and how these ungulates make foraging decisions at fine and broad scales. In Part 2, you'll examine how bark beetle outbreaks depend on and influence landscape pattern. And last, in Part 3, you'll analyze an aerial photograph of a fall webworm outbreak to determine some of the controls on the spatial configuration of outbreaks. You will need pen and paper, a calculator, a ruler, and access to Excel spreadsheet software to conduct some simple statistical analyses. For Part 3, print a copy of Figure 17.1 and Table 17.5 from the CD.

INTRODUCTION

Landscape structure refers to the physical elements within a landscape and their spatial relationship to one another. For example, fire in Yellowstone National Park (YNP) alters landscape structure by altering forests, creating patches of different size and shape where trees are absent. The spatial juxtaposition of good grazing sites across YNP represents an example of landscape structure. **Ecosystem processes** refer to ecological processes such as nutrient cycling and energy flow. For example, fire influences ecosystem processes by making some nutrients more readily available (in the form of ash) and reducing the availability of others (as they are volatilized by high temperatures); thus influencing rates of nutrient cycling.

Clearly, landscape structure can influence organisms in several ways, such as by facilitating or impeding the movement of animals from one location to another (see Chapter 13, Interpreting Landscape Patterns from Organism-Based Perspectives, and Chapter 15, Landscape Connectivity and Metapopulation Dynamics). For example, edges between burned, treeless areas and areas where trees are present become a critical part of the structure of the landscape as certain organisms are attracted to these edge environments (e.g., elk). Furthermore, organisms may also create landscape structure. Bison in Yellowstone National Park, for example, create shallow depressions, or wallows, in which they dust bathe (Meagher, 1973). These depressions are used as long as they are free of vegetation and do not hold water. Thus, organisms can both respond to and create landscape structure.

In responding to and creating new landscape structure, organisms can also alter the rate of ecosystem processes. For example, bison can move nutrients from one place to another by consuming grass in one patch in Yellowstone National Park, and then walking several kilometers to a better grazing patch before defecating. Furthermore, referring back to the example of wallow creation by bison, if the soil in a wallow is clayey in texture, repeated wallowing causes soil compaction, and the depression will accumulate water, becoming a shallow pond. This affects both nutrient availability and accumulation through alterations in soil pH and soil redox potentials (e.g., nitrogen availability declines in more acidic soils and microbial transformations of nutrients differ between aerobic and anaerobic soil conditions). Prairie dogs affect nutrient cycling and energy flow in grasslands by digging, bringing nutrient-rich subsurface soil to the surface where plant roots can access it. Energy flow is also affected by the heavy grazing these animals impose on plants, reducing their stature and affecting photosynthesis.

Many of the organisms mentioned are considered to be **ecosystem engineers** (Jones and Lawton, 1995). According to Jones et al. (1997),

> Physical ecosystem engineers are organisms that directly or indirectly control the availability of resources to other organisms by causing physical state changes in biotic or abiotic materials. Physical ecosystem engineering by organisms is the physical modification, maintenance or creation of habitats. The ecological effects of engineering on other species occur because the physical state changes directly or indirectly control resources used by these other species.

Thus, while the physical structure of the environment affects the survival of organisms, the actions of the "engineer" organism may simultaneously influence the structure of the environment.

A critical component of the interactions among organisms, ecosystem processes, and landscape structure is when feedbacks occurs between the components. A classic example of a system with **positive feedbacks** between organisms and ecosystems processes would be found where grazing organisms have created what is known as a grazing lawn (McNaughton, 1984). In this case, ungulates graze graminoids (grasses and grasslike plants) to a low height and maintain that low height with continued regrazing. Concurrently, urine and feces from the ungulates fertilize the site, and as a result, grasses flourish in this high-nutrient, high-grazing environment. The growth structure of the grasses also facilitates ungulate grazing because many leaves are compressed into a small volume of space. Thus, each ungulate bite removes a high amount of forage, optimizing ungulate forage intake. These lawns then become grazing "hot spots" with high rates of nutrient cycling and energy flow to ungulate herbivores. As a result, ungulates continually return to these locations to forage, leaving other sites nearly ungrazed. Thus, both the ungulate behavior and the ecosystem's responses have acted to increase ungulate grazing in a single location, a positive feedback. Conversely, when beavers cut down all of the trees adjacent to a pond, a **negative feedback** occurs. Without an additional food source, the beavers move away, the dam and the pond deteriorate, and this portion of landscape heterogeneity (the flooded area) is no longer maintained.

Researchers most commonly study herbivores when trying to understand the interaction between organisms and ecosystem processes. Well-studied examples include ungulate herbivores in Yellowstone National Park, beavers in aspen glades, bark beetles in pine forests, and prairie dogs in grasslands (Knight, 1987; Whicker and Detling, 1988; Turner et al., 1993; Pastor et al., 1996). In this lab, you will analyze several case studies of interactions between organisms and ecosystem processes.

As you read through the case studies, take detailed notes using the following steps to enable you to conduct a successful analysis of each study.

1. List the aspects of landscape structure that affect organisms living in the system. Some aspects may be explicitly discussed in the case study, whereas others will be ideas that you generate.

2. Construct a list of the activities of the organism(s) that affect the structure of the landscape.

3. Determine if positive or negative feedbacks occur between organisms and landscape structure and/or ecosystem processes.

4. After identifying feedbacks, determine what ecosystem processes are mediating these interactions. Is the rate of nutrient cycling, or the type of nutrients, being affected by the interactions occurring? Is energy flow through the system slowed, accelerated, or unaffected? Construct hypotheses about how the various interactions described could have effects on different nutrient cycles. List what types of data would be needed to test your hypotheses.

5. Note how the authors in the original papers conducted their research. How did they collect their data? How were those data analyzed?

To help you address these questions, a series of synthesis questions is given following each case study.

EXERCISES

Part 1. Ungulate Dynamics and Landscape Pattern in Northern Yellowstone National Park

Yellowstone National Park, located in the northern Rocky Mountains, contains the largest herds of native migratory ungulates in the coterminous United States. The dynamics of elk and bison herds have been studied extensively for decades, and the management of the animals and their habitat has long been controversial. In recent decades, elk population sizes in the northern range of Yellowstone have varied between a low of a few thousand animals in the 1960s, when population control measures were implemented, to over 18,000 by the summer of 1988. After the extremely large fires in the park during the summer of 1988, there was considerable interest in determining how the ungulate populations would respond (Christensen et al., 1989). These fires affected approximately 45% of the Yellowstone landscape, including 22% of the northern range. In this case study, we examine several ways in which fire has influenced the interactions between the ungulate herbivores (elk and bison) and their food resources. Two main themes are addressed, the effect of fire pattern and winter severity on: (1) ungulate survival and (2) foraging patterns.

EXERCISE 1.1
How Does Fire Pattern (Size and Spatial Arrangement) Affect Ungulate Winter Survival?

The interaction of fire, vegetation, and ungulate dynamics is an important management issue in Yellowstone National Park. Winter is a critical time for the survival of elk and bison because the availability of forage, which is often buried beneath snow, is limiting to their survival. Furthermore, the depth and density of the snow influences the ability of elk and bison to dig in the snow to access forage. Compounded with the effects of snow is the pattern of the 1988 fires. While burned areas provided less forage in the *first* postfire winter, burned areas provided *more* forage than unburned areas several years after the fire. If fire pattern is random, it might be difficult for ungulates to locate these high-resource areas in the years following a fire. However, if burned patches are clumped, these high-forage areas would be easier to locate. Thus, understanding the relative importance of fire pattern compared to other factors and their effects on ungulate survival is critical for ungulate management in the park. Due to the infeasibility of experimentally burning a large area of the park and manipulating the severity of winters, a modeling approach was employed.

THE NOYELP MODEL

The NOYELP (Northern Yellowstone Park) model (Turner et al., 1994) was constructed as a spatially explicit, individual-based model to reflect (1) the structure of the landscape, (2) the distribution of available forage, (3) winter severity on each pixel, (4) ungulate energetics and movement, and (5) ungulate winter survival. The model landscape was a grid of 1-hectare pixels representing an area of the winter range over 77,000 hectares. Realistic movement rules for the ungulates were employed while the model monitored their energy intake and expenditure to determine winter survival.

The model, commissioned by the National Park Service and the National Forest Service, was designed to answer the following questions:

1. Do large fires (such as the 1988 fires) affect the Yellowstone Ecosystem differently than do the small, widely dispersed fires that have occurred in recent historical times?

2. If future fire management were to include prescribed fire, what size patches should be burned by managers in this ecosystem?

Some of the additional hypotheses that were tested during model development included:

1. Are there threshold responses in this system?
   (a) Is there a fire size above which ungulate survival is greatly reduced?

  (b) Is there a population size for an ungulate above which survival
      is greatly reduced?

2. If thresholds exist, how does winter severity interact with thresholds
   in fire and population size?

Next, we will examine the basic elements of the model; however, more
detailed information can be found in Turner et al. (1994).

MODEL INPUT

Landscape Structure. Previously collected data on vegetation types and their
distribution (Despain, 1990) were simplified into 12 different types, including dry, mesic, moist, and wet grasslands; conifer forests; and aspen stands in
both burned and unburned condition. Topography (slope, elevation, and aspect) was also determined.

Distribution of Forage. Field data on forage biomass and quality were collected in the fall, immediately prior to winter snowfall, in both burned and
unburned areas of different habitat types. Forage availability in the model was
thus a function of this initial measurement of forage biomass and quality,
modified by the distribution of snow and the amount of grazing that occurred
in each pixel throughout the simulations.

Winter Severity. Winter severity was measured in terms of snow depth and
snow water content. Snow accumulation across the landscape was determined
using previously collected data from flat areas in the landscape that were then
modified using multipliers to portray snow accumulation on different slope
and aspect classes (Table 17.1).

Search and Movement Rules. Movement rules were developed on a pixel-by-
pixel basis for both elk and bison. Animals were able to "search" the pixel
in which they were placed during model initialization and would then move
in a probabilistic manner depending on pixel characteristics. If forage biomass
and availability dropped below a threshold level, the animals' probability of
moving out of that pixel would increase. Movement distances were based on
field records of animal movement patterns (Meagher, 1973; Houston 1982)
that suggest the maximum daily movement was 4 kilometers for elk and 2
kilometers for bison. It was also assumed that uphill movement required more
energy than movement on flat areas or downhill. Of course, the ability of the
animal to move was also important; if the animal lacked sufficient energy reserves to move (as determined by maintaining at least 75% of lean body
weight), then the animal would remain in the pixel and die.

MODEL OUTPUT

Ungulate Energetics. Daily energy balances were calculated for cows, calves,
and bulls of both ungulate species. Daily energy balances were a sum of energy gain (through forage energy content) minus losses (due to standing, rest-

TABLE 17.1
EFFECTS OF SLOPE AND ASPECT ON SNOW DEPTH. SNOW DEPTH
REPORTED ON A FLAT SURFACE (THE SNOW COURSE AT LUPINE CREEK)
WAS MULTIPLIED BY THE MULTIPLIER TO YIELD SNOW DEPTH ON
DIFFERENT SLOPE AND ASPECT CLASSES. ADAPTED FROM WALLACE
ET AL. (1993) AND TURNER ET AL. (1994).

Aspect Class	Slope (Degrees)	Multiplier
Nonforest and Burned Forest		
All	0–15	1.0
E, SE, W, SW	All	1.0
S	15–30	0.7
	>30	0.0
N, NW, NE	15–30	1.2
	>30	1.4
Unburned Forest		
All		0.9 * nonforest

ing, grazing, ruminating, thermoregulating, and movement, see Turner et al., 1994, for details).

Ungulate Survival. Daily and total survival was computed for both elk and bison. Daily survival was determined for individuals, whereas total survival refers to the total number of individuals surviving winter.

Actual data on winter severity (snow depth and snow water content) and ungulate population size were used to parameterize the model. The model was then run and the output (ungulate survival) was compared to what actually happened during those years. The data indicate that the model was, indeed, quite realistic in its portrayal of the northern range of Yellowstone Park (Table 17.2).

Question 1.1. Consider the questions that the NOYELP model was originally designed to answer (listed previously on page 253). Given that you can manipulate any of the input variables in any combination, how would you design a simulation experiment to answer these questions? Pick one of the questions and explain why you chose the design(s) that you did. Remember that the model was developed as a probabilistic model so that each model run is independent of the previous run. Therefore, you would run each experiment multiple times and collect model output from each run.

TABLE 17.2
VALIDATION OF THE NOYELP MODEL OUTPUT PREDICTIONS FOR BOTH
ELK AND BISON MORTALITY. ADAPTED FROM TURNER ET AL. (1994).

Year	Fire Status	Winter Severity	Elk Mortality (%)		Bison Mortality (%)	
			Observed	Simulated	Observed	Simulated
1987–88	Unburned	Mild	<5	0	0	0
1988–89	Postfire winter	Normal	38–43	40	20	18
1990–91	Later post-fire winter	Mild	<5	4	No data	0

EXERCISE 1.2
At What Spatial Scale Do Ungulates Make Foraging Decisions?

Broad-Scale Foraging Decisions

Observations of the northern range herds indicate that ungulates do not forage randomly across the entire landscape (Meagher, 1973; Houston, 1982; Pearson et al., 1995). Areas of deep snow are avoided, whereas sedge meadows and low-elevation, drier sites are preferred (Meagher, 1973; Pearson et al., 1995). Additional field observations of landscape use were collected using a series of viewing sites throughout the winter range (Table 17.3). These sites were visited biweekly during the winter and actively grazing animals or evidence of animal grazing was determined using binoculars and spotting scopes. Evidence of grazing is observable in the winter because elk paw through the snow creating craters in which forage is exposed. Similarly, bison use their heads as giant "brooms" to sweep patches clear of snow. The proportion of each site grazed at a particular time was mapped, and the proportions grazed over time were summed to create a map of values termed the minimum cumulative grazing index (MCGI). Table 17.4 lists statistically significant correlations of broad-scale landscape features with the MCGI using a Spearman rank correlation. Note that some correlations are negative and some are positive.

Fine-Scale Foraging Decisions

Within the preferred grazing areas discussed earlier, how do ungulates make foraging decisions? During the winter following the 1988 fires, the locations of feeding craters were mapped within a 30 × 30-meter area of the preferred grazing sites (Wallace et al., 1993). The craters were found to be randomly distributed within this area. The distribution of forage biomass

TABLE 17.3
Habitat Categories and the Proportion of Their Occurrence on the Northern Range. The Percentage of Sampling Area Is the Percentage of Each Habitat Type as It Occurred in the Visual Sampling areas. Values Listed under Northern Range Are the Percentage of that Habitat Type of the Entire Northern Range Landscape. Adapted from Pearson et al. (1995).

Habitat Type	Burned/ Unburned	% of Sampling Area	% of Entire Northern Range
Wet grassland	B	0.8	0.6
	U	8.2	12.8
Moist grassland	B	6.0	18.2
	U	48.4	51.7
Mesic grassland	B	9.2	5.7
	U	21.8	19.3
Dry grassland	B	0.7	0.7
	U	0.6	0.8

TABLE 17.4(a)
Correlation of Landscape Features (Spearman Rank Correlation Coefficients) with Minimum Cumulative Grazing Intensity (MCGI), an Indicator of the Proportion of an Area Grazed by Elk and Bison over Time. Adapted from Pearson et al. (1995). *NOTE:* See File Table **17.4(b).pdf** on the CD-ROM for a Description of the Variables Necessary to Interpret These Correlation Coefficients.

Variable	Correlation (r) with MCGI	
	1991 ($n = 4048$)	1992 ($n = 2980$)
Elevation	−0.12	−0.08
Mean annual precipitation	0.06	−0.12
Grassland habitat type	−0.22	−0.16
Slope	0.17	ns*
Aspect	ns	ns
Burn status	0.12	ns

*Not significant.

was also mapped in similar 30×30-meter areas and determined to be randomly distributed. Next, the ability of an ungulate to obtain forage in these areas was compared by simulating alternative random and nonrandom foraging patterns (Wallace et al., 1993). While a "smart" ungulate would forage in the highest biomass areas, the ungulates actually foraged randomly within these areas, resulting in approximately only *half* of the availability biomass being procured. Thus, ungulates were not able to predict the location of the best foraging spots at these fine scales, but rather appeared to be making their foraging choices at broader scales (Pearson et al., 1995).

Question 1.2. With which landscape feature is grazing (MCGI) most correlated? Is this a positive or negative correlation? Why do you think grazing is correlated most strongly with this feature and less so with others?

Question 1.3. What examples of negative feedbacks do you find in Exercise 1.2?

Question 1.4. Is the spatial distribution of nutrients in this ecosystem being affected by the interactions occurring?

Reorder: Are the interactions occurring in this ecosystem affecting the spatial distribution of nutrients?

Question 1.5. Is energy flow through the system slowed, accelerated, or unaffected by ungulate interactions with the landscape?

Question 1.6. The model and results discussed are very specific to Yellowstone National Park. What types of information would you need to obtain before using this model in other systems in which elk and bison are present?

Question 1.7. Foraging decisions appear to be made by these ungulate species at a particular spatial scale (see Senft et al., 1987). How do you think the results would differ if elk and bison made their foraging decisions at a coarser scale? An even finer scale?

Question 1.8. The model describes ungulate responses during the winter when there is no plant regrowth. How would you modify this model to deal with grazing during the growing season?

Part 2. Interactions between Bark Beetle Outbreaks and the Forest Landscape Mosaic

Herbivorous insect outbreaks represent episodic disturbances capable of dramatically altering landscape structure and ecosystem function (Rykiel et al., 1988). Bark beetles, a general term for several members of the family Scolytidae, feed on the subcortical region (under the bark) of specific host tree species, causing damage and mortality over potentially large areas (Thatcher et al., 1982). Bark beetles share several important characteristics with other outbreak insects, and understanding these life-history characteristics is fundamental to determining the spatial patterns of outbreaks.

An important feature of bark beetle biology is their ability to persist in small numbers until conditions (usually weather) are favorable for reproduction (Edwards and Wratten, 1980). Under the right circumstances, bark beetles can then increase tenfold in a single generation (Rykiel et al., 1988), quickly reaching epidemic levels. Another critical feature of bark beetle biology is their ability to move across the landscape in search of suitable hosts. For example, the southern pine bark beetle (*Dendroctonus frontalis*) requires mature pines (*Pinus* spp.) for successful colonization. Furthermore, bark beetles preferentially choose trees whose defenses are diminished by stress. Away from a favorable host, beetle survival decreases sharply with time, due in part to predation, but primarily because of a low tolerance for temperature extremes.

In large part these features of bark beetle life history determine spatial patterns of current and future bark beetle outbreaks. For example, southern pine bark beetles can decimate large patches of suitable forest (in this case, mature pine). If the outbreak begins in or near a large stand of mature pines, the epidemic may spread over thousands of hectares until unsuitable habitat is reached or weather conditions change. However, if a beetle population explosion occurs in a small area of suitable habitat surrounded by a large area of unsuitable habitat such as young pine, the epidemic is confined to a small area, essentially because few beetles can survive the long journey to distant suitable hosts. Over time, beetle-induced tree mortality can create a landscape mosaic of different aged stands in different sized patches.

Question 2.1. Does the relationship between patterns of bark beetle outbreaks and the landscape mosaic represent a positive or negative feedback? Explain your answer.

Question 2.2. How might ecosystem processes mediate the bark beetle/landscape interaction?

Question 2.3. In the previous case study, Turner et al. (1993) used a spatially explicit model of elk and bison foraging to predict winter survival in Yellowstone National Park. The model relies heavily on rules of movement for large herbivores developed from field observations. What are some rules of movement for the spread of bark beetle infestations? (*HINT*: Think both in the practical sense and in the theoretical sense, considering Part 1 of Chapter 15, Landscape Connectivity and Metapopulation Dynamics, and Chapter 13, Interpreting Landscape Patterns from Organism-Based Perspectives.)

Question 2.4. How might bark beetle survival be affected by the homogenization of the landscape that occurs with plantation forests?

Part 3. An Analysis of Landscape Usage Patterns by the Fall Webworm

Fall webworms (*Hyphantria cunea*), a generalist lepidopteran herbivore (member of the butterfly and moth group), provide another example of insects whose outbreaks interact with landscape structure. The larvae of the *Hyphantria* moth form silk tents over the foliage of deciduous trees, which they consume until reaching maturity. Work with similar lepidopteran herbivores has shown that many species choose locations in full sun over those in shaded environments (Louda and Rodman, 1996). Also, the severity of tent-forming caterpillar outbreaks is known to increase with forest fragmentation (Roland, 1993).

In the summer and fall of 1996, central Oklahoma experienced a severe outbreak of the fall webworm. During this time Wallace et al. (in prep) performed a study to determine how forest structure influenced webworm distributions on the landscape shown in the aerial photograph in Figure 17.1. The locations of webworm-infested trees are identified with white circles. Edges of the forested areas and forested openings appear as lighter and gray while forested areas are darker gray and have a more "textured" appearance. You will conduct two different spatial analyses using the data collected in that study.

FIGURE 17.1
Black-and-white aerial photograph showing fall webworm distributions in a grassland–forest mosaic in Oklahoma. Forested areas are dark, and open, nonforested areas are lighter gray. Trees in which webworms were present are marked with white circles. This image has also been included on the CD-ROM under the directory for this lab and is entitled **webs.pdf.** The file can be viewed and then printed using Adobe Acrobat Reader.

EXERCISE 3.1
Do Webworms Preferentially Choose Forest Edge Trees?

Here you will conduct a *t*-test to determine if infested trees are closer to edges and openings than would be expected by chance. You will compare the mean distance from a random point to the nearest infested tree to the mean distance between and infested tree and a forest edge or opening.

1. First, use a ruler to demarcate axes for an *x-y*-coordinate system along the two edges of Figure 17.1.

2. Use a random numbers table (or a calculator with a random number generator) to locate a random point in the photograph: generate a random number for the *x*-coordinate, and then another random number for the *y*-coordinate. Record your *x*- and *y*-coordinates in Table 17.5 which can be printed from the CD.

3. Using the ruler and the *x*- and *y*-coordinates, locate your first random point on the photograph.

4. Determine the distance to the nearest infested tree from this random point. Mark this infected tree on the image and enter the distance in Table 17.5.

5. Next, record the distance from the nearest infested tree to the nearest forest edge or opening.

6. Repeat steps 2–5 until you have 15 observations.

Next, you'll conduct a *t*-test (identical to the tests you did in Part 1 of Chapter 11, Landscape Disturbance: Location, Pattern, and Dynamics).

1. Enter the two columns of distance data from Table 17.5 into a worksheet in Excel, including the column headings.

2. Under the **Tools** menu, select **Data Analysis.** In the **Analysis Tools** window, scroll down to "***t*-test: Two-Sample Assuming Unequal Variances.**" (*NOTE:* If Data Analysis does not appear under the **Tools** menu, select **Add-Ins**, and add the **Analysis Tool Pak.**)

3. Enter the "distance to nearest infested tree" and "distance to nearest forest edge" columns into the variable 1 and variable 2 range boxes by highlighting the data in the worksheet using the mouse.

4. In the **Output Options** box, select the circle next to **Output Range,** and enter a destination cell or range of cells for the *t*-test output in the worksheet, usually a cell a few columns away from your data.

5. Click **OK**, and the *t*-test will be computed.

If the absolute value of the *t*-statistic computed for your data is greater than the *t*-critical value provided in the output, you will reject the null hypothesis, but accept the alternative hypothesis that the mean distance between infested trees and edge is less than or greater than would be expected by chance.

EXERCISE 3.2
Are Webworm-Infested Trees Clumped Throughout the Landscape?

A second analysis can be made using the aerial photograph to determine whether fall webworm-infested trees exhibit a clumped distribution. This can be determined using the T-square index of spatial pattern, termed C (Ludwig and Reynolds, 1988).

1. From the randomly selected points used in the previous analysis, determine the distance from that point to the nearest infested tree (x_i). (You should already have this number from your previous *t*-test analysis.)
2. From that tree (x_i), determine the distance to the nearest infested neighbor (y_i).
3. Use the following formula to calculate C where n = the total number of sample points.

$$C = \frac{\sum \dfrac{x_i^2}{(x_i^2 + \frac{1}{2} y_i^2)}}{n}$$

4. Determine the value of C. C is approximately 0.5 for random patterns, significantly less than 0.5 for uniform patterns, and significantly greater than 0.5 for clumped patterns.
5. To determine if the value of C is significantly different from 0.5, use the following *z*-test:

$$z = \frac{C - 0.5}{[1/(12n)]^{0.5}}$$

If z is greater than or equal to 1.96, then C is significantly different from half, or in other words, the webworm-infested trees are nonrandomly distributed on the landscape.

Question 3.1. Although tree mortality from webworm infestation is rare, indirect influences on host health such as increased disease susceptibility may be important for long-term host survival. Paying particular attention to forest fragmentation, discuss the possible interactions between webworm distribution and landscape structure. How might this interaction affect ecosystem processes? (HINT: see Turner et al., 1989.)

Question 3.2. Webworm outbreaks may cause widespread damage to trees and mast crops (fruits and nuts). White-tailed deer (*Oidocoileus virginianus*) populations in central Oklahoma depend heavily on mast production for winter forage. How might patterns of webworm damage influence patterns of deer survival? Draw a conceptual model of this interaction.

Question 3.3. Unlike pine bark beetles, fall webworms are known to use a large array of host species (Nothnagle and Schultz, 1987). However, some species, such as pecan (*Carya illinoensis*) and sweetgum (*Liquidambar styraciflua*), are chosen preferentially as hosts. In addition, some unpalatable tree species such as red cedar (*Juniperus virginiana*) are seldom (if ever) used by fall webworms. If individual trees of the preferred species were protected from webworm infestation when located in the midst of unpalatable species, how might this so-called "plant defense guild" (Atsatt and O'Dowd, 1976) influence landscape patterns of webworm infestation? How could you test this hypothesis?

CONCLUSIONS

In this lab you have seen and used different techniques for analyses of the effects of landscape structure and ecosystem function on organisms as well as the feedbacks that organisms have on the structure and function of ecosystems and landscapes. Both modeling and direct observation of organism distribution and landscape structure and function are important tools in this endeavor. There are numerous other examples in the literature; however, most work of this sort is conducted in the field (or in a model parameterized with field data) rather than in the laboratory. Nevertheless, laboratory data can also be useful to understand some of the finer-scale mechanisms that may be driving a particular system. While it may seem intuitively obvious that organisms can influence ecosystem processes and landscape structure, much work is still needed to understand the mechanisms behind these effects.

BIBLIOGRAPHY

Note. An asterisk preceding the entry indicates that it is a suggested reading.

*ATSATT, P. R., AND D. J. O'DOWD. 1976. Plant defense guilds. *Science* 193:24–29. This benchmark paper in ecology, the first to describe a potentially mutualistic interaction among plants involving interaction with a separate trophic level, represents one of the finest early studies of spatial ecology, as well.

CHRISTENSEN, N. L., J. K. AGREE, P, F. BRUSSARD, J. HUGHES, D. H. KNIGHT, G. W. MINSHALL, J. M. PEEK, S. J. PYNE, F. J. SWANSON, J. W. THOMAS, S. WELLS, S. E. WILLIAMS, AND H. A. WRIGHT. 1989. Interpreting the Yellowstone fires of 1988. *BioScience* 39:678–685.

Despain, D. 1990. *Yellowstone Vegetation.* Roberts Reinhart, Boulder, Colorado.

EDWARDS, P. J., AND S. D. WRATTEN. 1980. *Ecology of Insect-Plant Interactions.* Edward Arnold, London.

Houston, D. G. 1982. *The Northern Yellowstone Elk.* MacMillan, New York.

*JONES, C. G., AND J. H. LAWTON. 1995. *Linking Species and Ecosystems.* Chapman and Hall, New York. This seminal volume on linking processes across scales includes chapters on linkages between species-level processes and landscape and ecology levels with examples of population structure and function, evolution, and trophic-level interactions.

JONES, C. G., J. H. LAWTON, AND M. SHACHAK. 1997. Positive and negative effects of organisms as physical ecosystem engineers. *Ecology* 78:1946–1957.

KNIGHT, D. H. 1987. Parasites, lightning and vegetation mosaic in wilderness landscapes. In M. G. Turner, ed. *Landscape Heterogeneity and Disturbance*. Springer-Verlag, New York, pp. 59–83.

LOUDA, S. M., AND J. E. RODMAN. 1996. Insect herbivory as a major factor in the shade distribution of a native crucifer (*Cardamine cordifolia* A. Gray, bittercress). *Journal of Ecology* 84:299–237.

LUDWIG, J. A., AND J. F. REYNOLDS. 1988. *Statistical Ecology: A Primer on Methods and Computing*. John Wiley and Sons, New York.

MCNAUGHTON, S. J. 1984. Grazing lawns: Animals in herds, plant form and coevolution. *American Naturalist* 124:863–886.

MEAGHER, M. M. 1973. The bison of Yellowstone National Park. National Park Service Scientific Monographs, No. 1.

NOTHNAGLE, P. J., AND J. C. SCHULTZ. 1987. What is a forest pest? In P. Barbosa and J. C. Schultz, eds. *Insect Outbreaks*. Academic Press, New York, pp. 59–80.

PASTOR, J., A. DOWNING, H. E. ERICKSON. 1996. Species-area curves and diversity-productivity relationships in beaver meadows of Voyageurs National Park, Minnesota, USA. Oikos 77:399–406.

PEARSON, S. M., M. G. TURNER, L. L. WALLACE, AND W. H. ROMME. 1995. Winter habitat use by large ungulates following fire in northern Yellowstone National Park. *Ecological Applications* 5:744–755

ROLAND, J. 1993. Large-scale forest fragmentation increases the duration of tent caterpillar outbreak. *Oecologia* 93:25–30.

RYKIEL, E. J., R. N. COULSON, P. J. H. SHARPE, T. F. H. ALLEN, AND R. O. FLAMM. 1988. Disturbance propagation by bark beetles as an episodic landscape phenomenon. *Landscape Ecology*. 1:129–139.

*SENFT, R. L., M. B. COUGHENOUR, D. W. BAILEY, L. R. RITTENHOUSE, O. E. SALA, AND D. M. SWIFT. 1987. Large herbivore foraging and ecological hierarchies. *BioScience* 37:789–799. This paper will also be a benchmark paper in ecology because of its efforts to link foraging processes across scales. Although few people now use the concept of overmatching as described in the paper, many use the conceptual processes developed here.

THATCHER, R. D., J. L. SEARCHY, J. E. COSTER, AND G. D. HERTEL. 1982. The southern pine beetle. USDA Forest Service Technical Bulletin 1631.

*TURNER, M.G., R. H. GARDNER, V. H. DALE, AND R. V. O'NEILL. 1989. Predicting the spread of disturbance across heterogeneous landscapes. *Oikos* 55:121–129. This paper describes, in both mathematical and ecological terms, how heterogeneity should influence the spread of disturbances, populations, matter, and information across a landscape. It also describes a null-model approach for testing the effects of heterogeneity on these flows.

TURNER, M. G., Y. WU, W. H. ROMME, AND L. L. WALLACE. 1993. A landscape simulation model of winter foraging by large ungulates. *Ecological Modeling* 69:163–184.

TURNER, M. G., Y. WU, L. L. WALLACE, W. H. ROMME, AND A. BRENKERT. 1994. Simulating interactions among ungulates, vegetation, and fire in northern Yellowstone park during winter. *Ecological Applications* 4:472–496.

WALLACE, L. L., AND S. T. GRAY. In Prep. Impact of a fall webworm (Hyphantria cunea) outbreak on prairie woodlands.

WALLACE, L. L., M. G. TURNER, W. H. ROMME, R. V. O'NEILL, AND Y. WU. 1993. Scale of heterogeneity of forage production and winter foraging by elk and bison. *Landscape Ecology* 10:75–83.

WHICKER, A. D., AND J. K. DETLING. 1988. Ecological consequences of prairie dog disturbances. *BioScience* 38:778–785.

CHAPTER 18

MODELING ECOSYSTEM PROCESSES

SARAH E. GERGEL AND TARA REED-ANDERSEN

OBJECTIVES

Understanding and predicting rates of ecosystem processes (e.g., soil erosion, nutrient flux) across large heterogeneous landscapes is an emerging challenge in ecosystem and landscape ecology. Also, many current problems in ecosystem management (e.g., maintenance of water quality and reduction of soil erosion) occur over broad spatial scales and across ecosystem boundaries. Historically, ecosystem ecologists and watershed hydrologists have often used fine-scale plot experiments to infer rates of ecosystem processes at broader scales. This can present difficulties, as the results of fine-scale studies may not reflect the heterogeneity of a large area. Because collection of ecosystem data at broad scales is often difficult and costly, modeling is a vital tool for addressing both basic and applied questions in this area. In this lab you will examine several fundamental issues of modeling landscape-level ecosystem processes in order to

1. gain an appreciation for the need to examine ecosystem processes at broad scales;
2. learn to conceptualize how ecosystem processes can be modeled at the scale of a landscape; and
3. examine the implications of heterogeneity in rates of ecosystem processes.

In this lab we focus on modeling phosphorus flux through an agricultural watershed. This exercise will involve the examination and understanding of a simple landscape ecosystem model and then provide you with the conceptual and technical tools to design an ecosystem model of your own. This exercise includes an Excel spreadsheet file, **ecosys.xls,** which will constitute your modeling environment.

NOTE: Save an extra copy of the model that you *do not* manipulate as a backup in case you accidentally irreversibly alter the model.

INTRODUCTION

Eutrophication, or the enrichment of aquatic systems by excessive input of nutrients, constitutes the major threat to water quality in the United States (U.S. Environmental Protection Agency, 1997). Phosphorus (P) is often the limiting nutrient to algal productivity in freshwater systems. As a result, P enrichment can lead to toxic algal blooms and increases in hazardous protozoa (Schindler, 1977), which can threaten fisheries, drinking water supplies, and recreational opportunities (Carpenter et al., 1998). Nuisance algal blooms can also reduce habitat diversity in shallow waters and deplete oxygen in bottom waters causing massive fish die-offs (Kaufman, 1993). Additionally, the water may smell and taste foul and cause skin irritation.

The most ubiquitous cause of eutrophication is nonpoint source pollution (U.S. Environmental Protection Agency 1990). **Nonpoint source pollution** refers to material entering aquatic systems from diffuse sources, such as runoff from agricultural fields. This is in contrast to **point sources,** such as sewage treatment outflow pipes. Agricultural areas, particularly during storm events, contribute significantly to nonpoint source phosphorus pollution (Osborne and Wiley, 1988; Omernick et al., 1991; Correll et al., 1999). **Riparian buffer strips,** bands of uncultivated vegetation adjacent to surface waters, slow P flow and can be used to mitigate fertilization of water bodies in agricultural areas (Peterjohn and Correll, 1984; Gregory et al., 1991; Osborne and Kovacic, 1993).

In this lab we present an ecosystem model of the agricultural landscape surrounding a canal. The canal leads to a nearby lake that is used by the public for recreational swimming and boating. The model represents the flow of phosphorus from fertilized agricultural fields, through the riparian buffer strip, and into the canal during a storm. While the model presents a highly simplified version of P dynamics, it provides a useful introduction to modeling ecosystem processes at the scale of a landscape. This model is designed to address questions such as, *How much phosphorus can farmers apply to their fields without causing severe algal blooms?* and, *At a given phosphorus application level, how much phosphorus must be retained by the buffer strip to maintain low phosphorus levels in the canal?*

EXERCISES

Part 1. Conceptualizing Landscape-Level Ecosystem Models: Phosphorus Loading in an Agricultural Landscape

Model Description

Open the file **ecosys.xls** using Excel spreadsheet software. The spreadsheet has been configured to represent a model agricultural landscape. The brown cells on the landscape represent the farmed lands. After fertilizer is applied, some P flows downhill toward the canal, represented by blue cells. The green cells represent vegetated buffer strips. The number in each cell represents the total amount of P available to leave that cell, after within-cell uptake and processing is taken into account. Each cell in the model landscape represents one hectare (ha), a 10,000-m^2 area. This model approximates P flow across an agricultural landscape during a single storm event using the following simple parameters.

MODEL INPUT

Storm flow volume (m^3/ha) [G6] is the total amount of stream flow in each cell in the canal for the duration of the storm.

Buffer absorption capacity (kg/ha) [G10] represents the ability of the buffer strip to prevent the passage of P to the next cell, expressed as the total amount of P that could be retained by the buffer cell. Riparian buffers stop the flow of P in a variety of ways, including uptake by plants, trapping of soil to which the phosphorus is bound, and soil adsorption and immobilization. Here we combine all the mechanisms into one equation for simplicity, representing the sum total of the buffer strips' ability to prevent P from entering the canal. In our idealized landscape, values for this parameter range from 20 to 40 kg/ha (Peterjohn and Correll, 1984; Osborne and Kovacic, 1993).

Amount of phosphorus applied (kg/ha) [G15] refers to the amount of P in the fertilizer applied to each individual cell in the model. This model assumes that fertilizer is applied evenly throughout the field. In practice, the amount of P applied through fertilizer is highly variable, ranging from 50 to 200 kg/ha (Nowak et al., 1996).

Transfer Coefficient. Our model assumes that farmers are not applying fertilizer in the rain and that 60% of the P in the cell runs off one pixel to another during a storm. This percentage is a simplification; in a real agricultural field the amount of P runoff would vary with vegetation, slope, and rainfall intensity.

MODEL OUTPUT

Total phosphorus loading (kg) [G19] is the sum total of P entering the canal waters.

In-stream phosphorus concentration (mg/m^3) [G22] is the resulting concentration of P in the canal surface water after the total P is thoroughly mixed throughout the water column. The total concentration was multiplied by 1,000,000 to convert kg to mg. We then divided by the number of cells in the stream (75) times storm flow to find the mg/m^3. When the in-stream P concentration exceeds 75 mg/m^3, the system is at risk for algal blooms (Lathrop et al., 1998).

Exploring the Model—How Does It Work?

Agricultural Field. The brown cells on the spreadsheet represent farmed areas. These fields slope down to the stream running down the middle of the spreadsheet. The number in each cell represents the amount of P "left over" after uptake within the cell is accounted for; that is, the amount available to leave the cell and flow downhill to the next cell.

Select cell N4. Note the equation for the amount of P that leaves this cell. It is composed of two parts. The first part of the equation, M4 + $G$15, calculates the amount of P entering the cell. M4 is the amount flowing in from the adjacent upstream cell. $G$15 is the amount of fertilizer applied directly to the cell by the farmer (an input parameter you can alter). The sum of these numbers is the total amount that entered the cell. However, not all of this P flows to the adjacent downhill cell during a storm, as some is taken up by plants, adsorbed to soil particles, or leached into groundwater before it reaches cell O4. Thus, the 0.6 multiplier, or transfer coefficient, accounts for the fact that only 60% of the P that entered the cell can be washed into the next cell (i.e., 40% is taken up). This is an oversimplification. In reality, soil cannot bind an infinite amount of P. The model also assumes that flow is unidirectional, downhill toward the canal. This is another simplification. Flow is likely to be much more complex in a natural landscape.

Buffer Strips. Next examine the buffer strips (the green areas) along the banks of the canal. Select cell T22.

Question 1.1. Write the formula for cell T22, and explain in words what it means. *NOTE*: These cells contain the Excel equivalent of an "if–then" statement to prevent the program from printing negative numbers. These statements read: IF($x < y$, print this if true, print this if false). For your answer, describe what the equation in the "print this if false" section means.

Drainage Canal. Eventually some P may make its way into the canal. Notice the differences in loading values for near-shore stream cells due to the variable width of the buffer at different sites.

Question 1.2. Write the formula and explain in words how the output parameter "in-stream phosphorus concentration" is calculated.

Now, select cell V15. Enter a value of 5 into the cell. Repeat for cells V4, V10, V11, V12, and V22. Did the **in-stream phosphorus concentration** increase? By how much?

You just simulated several "cow patties" produced by a small herd of cows wading in the canal.

Part 2. Heterogeneity in Ecosystem Processes

In this section you will manipulate different components of the model to gain familiarity with how it can be used to explore alternative scenarios involving spatial variation in parameters and rates.

EXERCISE 2.1
Phosphorus Application Rates

As with all simulation models, important simplifying assumptions have been made for this model. Notice that all agricultural areas, for example, have the same amount of P applied to each cell. In reality, the amount applied to each cell could vary for several reasons. For example, a farmer might determine that a certain area of the field needs more fertilizer than other areas due to soil type. Also, different fertilizer application techniques might result in uneven P application throughout a watershed.

Consider that two farmers live on opposite sides of the creek and simulate the effect of different farming practices on the landscape. Implement this by changing the formulas in the cells, or by summing total P runoff for different sides of the landscape under alternative P application rates.

Question 2.1.
(a) Explain your modification.
(b) What effect does this heterogeneity in fertilizer application have on the in-stream P concentration?
(c) Another difference in P movement could be due to differences in crop type. For example, hay production requires less P than corn (Newman, 1997). Describe how you would change the model to incorporate differences in crop type. What equation would you change? How would you change that equation?

EXERCISE 2.2
Topographic Heterogeneity and Transfer Rates

Additional factors may cause heterogeneity in ecosystem process rates. Consider the importance of heterogeneity in the rates of P movement across the landscape caused by topography. Erosion of P-containing sediment is often greater in areas of steep slopes, particularly during rain events.

Change the model to account for slope differences throughout the landscape. The easiest way to do this is by changing the amount of P leaving an individual cell, thereby simulating a reduction or increase in the processing time for P in that cell. Right now the processing rate is 40% of the inputs (i.e., 60% exits the cell), but this might vary depending on whether the slope is gentle or steep.

Question 2.2. Describe the changes you made and the effects on P loading and concentration. What other factors might you expect to influence the movement of P (other than the transport across buffers)?

EXERCISE 2.3
Variation in Buffer Strip Width vs. Application Rates

You probably noticed earlier that the width of the buffer strip is important in determining P loading into the canal in this model. For the sake of managing water quality in the surrounding surface waters, a land manager or farmer may be interested in the relative importance of the width of the buffer strip versus the amount of fertilizer applied by a farmer in influencing total P inputs.

Question 2.3. For the modeling scenario examined here, does it appear that individual farmer behavior (i.e., application rates) or buffer width is more important in maintaining low concentrations of in-stream phosphorus? Answer in light of the constraints of the model and the range of parameters given.

Continue to manipulate the model, changing parameters at will. *Be certain that you understand all the model parameters and how all model formulas were derived.*

PART 3. CONSTRUCTING YOUR OWN LANDSCAPE MODEL

Now that you have been introduced to the fundamentals of a simple landscape model and have had a chance to explore its parameters and possibilities, you have the tools to design your own landscape model. You will use the same basic concept of combining cells of landscape elements in Excel to build your own landscape-level ecosystem model. Your instructor may assign either the "Basic Version" below, or the "Advanced Version" on page 274.

Basic Version

Your task is to build a model to answer a specific question regarding the dynamics of P runoff in an urban landscape. Your urban environment is a city, such as Chicago or Seattle. In Excel, you will model a city using a set of cells representing different elements of the urban environment (Table 18.1). Each element has its own level of P runoff and/or absorption. Using your imagination, create a city that contains at least a small proportion of all of the provided urban elements. Your city is adjacent to a small river that receives urban storm-water runoff.

TABLE 18.1
MODEL PARAMETER ESTIMATES FOR USE IN DESIGNING AN URBAN
LANDSCAPE. NUMBERS ARE BASED ON AND ADAPTED FROM BANNERMAN
ET AL. (1993), REUTER ET AL. (1992), AND OSBORNE AND KOVACIC
(1993).

Land-Cover Type	Amount of Phosphorus Produced (g/20 m^2)	Phosphorus Absorption Capacity (g/20 m^2)	Simplified Transfer Coefficient (proportion)
Lawn (heavily fertilized)	30	—	0.60
Lawn (slightly fertilized)	4	—	0.60
City park (slightly fertilized)	4	—	0.60
Residential homes	70	—	1
Apartments	30	—	1
Commercial district	20	—	1
Industrial district	40	—	1
Construction site	200	—	1
Runoff treatment wetland	—	50	0.60
Forest	—	40	0.60

Design and then manipulate your model specifically to answer at least *one* of following questions:

1. City parks can be sinks for P, although they are slightly fertilized. What proportion of the city must be occupied by parks to maintain in-stream P levels below 75 mg/m^3? How does the spatial arrangement of the parks affect the proportion of the city that parks must occupy to maintain in-stream P levels below 75 mg/m^3?

2. What proportion of the stream must be bordered by runoff treatment wetlands in order to reduce in-stream P concentrations by 10%? By 50%? To eliminate P input altogether? What proportion of the stream must be bordered by treatment wetlands to maintain in-stream P levels below 75 mg/m^3?

3. Keeping the total area in housing constant, what effect does varying the proportions of residential housing in apartments versus homes (e.g., 30/70, 50/50, 90/10) have on P runoff to the stream? What proportions would you recommend to maintain in-stream P levels below 75 mg/m^3?

Model Parameters. You are provided with parameter estimates in Table 18.1. Notice, however, that the resolution of the runoff and absorption estimates is different than for the model you examined in Parts 1 and 2. You will probably want to adjust the scale of your model from the 1-hectare resolution used in the agricultural model, as city lot sizes are rarely that large. Here, we have provided the model parameters in units of g/20 m². For an urban landscape, 20 m² cells roughly approximate the minimum size (or spatial grain) of the landscape elements you will be modeling. For example, you could combine one residential housing pixel with one lawn pixel to represent one residence.

To incorporate both urban and agricultural areas in your landscape, you can use values from Part 1, but you will need to do some conversions (remember 1 ha = 10,000 m²). You may adjust the grain size further as appropriate for your model and the questions you are trying to address, but be sure to choose an appropriate grain size for your model and adjust the runoff and absorption capacity values accordingly. Last, you can assume that all processes that contribute to P runoff and/or absorption have been taken into account with the parameters given.

Transfer Coefficients. In addition to the absorption capacity of a land-cover type, the amount of P transferred to the next cell may also be diminished by a transfer coefficient in areas that are sources of P. This reflects that some land-cover types are less permeable to runoff than others so that more of the runoff moves from one cell to the next. In the agricultural model we used a transfer coefficient of 0.6, meaning that only 60% of the P in a cell was available to move out to the next cell. In this section only wetland areas and forests have absorption capacity values. We have, however, included transfer coefficients to account for soil permeability in lawns and parks and other source areas in the spreadsheet, which we will examine now.

Building Your Model. Switch to the second page of the Excel file by clicking on the tab **Urban Landscape** on the bottom of the spreadsheet. Again, here are all the elements with which to build your urban landscape, identical to those in Table 18.1. Click on the cells in the Equations column to view the equations, incorporating transfer coefficients in some cases, for the different land-cover types. The cells in the Example column can be cut and pasted into the spreadsheet to build your urban landscape.

NOTE: These equations work for P flow only from left to right. Unless you want to rewrite some of the equations to represent flow in the opposite direction, place your river, stream, or canal on the right-hand boundary of your landscape. Be sure to examine each cell to see which other cells are referenced.

Construct your model in the same general form as the model in Part 1. For simplicity, you may assume that flow is unidirectional, downhill toward the canal. Thus, as before, the P values in each cell represent the amount leaving that cell. This includes the runoff entering from the adjacent upstream cell plus or minus the runoff/absorption estimate for that land-cover type, and in some cases, a transfer coefficient. *Remember that the number in each cell should represent the total P available to leave the cell, after any within-cell uptake or processing or reduction due to the transfer coefficient.* The concentration in the water can be calculated by the amount of P flowing into the stream cell multiplied by the total volume of water that flowed through the stream during the storm event. When your model construction and manipulation are finished, complete the model Write-Up portion of the lab.

Advanced Version

Your task is to model any landscape-level ecosystem process of your choosing. You will use the basic concept of landscape element blocks in Excel, but you are free to design those elements using your own knowledge, experience, and imagination. As in the basic version, *your model must be designed to answer a clearly defined question* (or set of related questions), but you will choose the question yourself. Be sure that you have a clear understanding of the underlying assumptions of your model throughout the building process, and be able to state those assumptions clearly. Be sure to determine explicitly the grain size of your model and whether the numbers in each cell represent amounts *entering* or *leaving* each cell. If you have more than one day to complete the assignment, we recommend that you spend some time researching the literature and use realistic parameters to construct your model. Keep in mind that you must be able to manipulate your model to address your initial question. When the model and manipulation are finished, complete the model Write-Up that follows.

Hints for modeling

1. Consider using the **Format**, then **Cell**, then **Patterns** commands on your spreadsheet's pull-down menu to assign different colors identifying different landscape elements.

2. Learn how to use $ symbols for cutting and pasting. For example, if you wanted to copy the formula =$F$6+5 from one cell to a cell in the next column over, $F preserves the column reference, while $6 preserves the row reference; thus, the formula would remain =$F$6+5 when copied and pasted. Otherwise, the formula typed as =F6+5 becomes =G6+5 when copied one to the right, or becomes =F7+5 when copied to the cell below.

WRITE-UP

Include the following sections in your report:

1. Introduction
   (a) State the question(s) your model addresses.
   (b) Provide some context for why this question is important.

2. Description of Model
   (a) State the underlying assumptions of your model.
   (b) Describe your model. Define the spatial and temporal scale of your model. (For the advanced version, list and explain all model parameters.)

3. Simulations and Results
   (a) Clearly describe each "simulation experiment" with the model and summarize the results.
   (b) Answer the question(s) your model was designed to address.
4. Discussion
   (a) What are the implications of heterogeneity in rates of ecosystem processes in your model scenario?
   (b) Within the realm of the ecosystem process that you have modeled, what are the limitations of your model? Why?
   (c) What additions/modifications would you make to your model to address the questions listed?
   (d) When would considering the spatial arrangement of landscape elements or the role of landscape heterogeneity not matter to your results?
   (e) When would sampling at broad scales not be important?
   (f) How would a broader temporal scale affect your results?
5. **Literature Cited** (not included in page limits)
6. **Appendix** (not included in page limits)
   (a) A copy of the answers to the exploratory questions posed in Parts 1 and 2 of this exercise
   (b) Printout of the Excel file containing *your* model

Your instructor will determine page lengths depending on the amount of time you have to complete your assignment. Consider giving oral presentations of your results.

BIBLIOGRAPHY

Note. An asterisk preceding the entry indicates that it is a suggested reading.

*Burke, I. C., D. S. Schimel, C. M. Yonker, W.J. Parton, and L. A. Joyce. 1990. Regional modeling of grassland biogeochemistry using GIS. *Landscape Ecology* 4(1):45–54. A good example at simulating the spatial variability in the storage and flux of carbon and nitrogen.

Bannerman, R. T., D. W. Owens, R. B. Dodds, and N. J. Hornewer. 1993. Sources of pollutants in Wisconsin stormwater. *Water Science and Technology* 28(3–5): 241–259.

Carpenter, S. R., N. F. Caraco, D. L. Correll, R. W. Howarth, A. N. Sharpley, and V. H. Smith. 1998. Non-point pollution of surface waters with phosphorus and nitrogen. *Ecological Applications* 8(3):559–568.

Correll, D. L., T. E. Jordan, and D. E. Weller. 1999. Effects of precipitation and air temperature on phosphorus fluxes from Rhode River watersheds. *Journal of Environmental Quality* 28:144–54.

*Fisher, S. G., N. B. Grimm, E. Marti, R. M. Holmes, and J. B. Jones Jr. 1998. Material spiraling in stream corridors: A telescoping ecosystem model. *Ecosystems* 1(1):19–34. A new look at the concept of nutrient spiraling in streams emphasizing the importance of spatial heterogeneity in influencing rates of nutrient processing.

GREGORY, S. V., F. J. SWANSON, W. ARTHUR, AND K. W. CUMMINS. 1991. An ecosystem perspective of riparian zones: Focus on the links between land and water. *Bioscience* 41(8):540–551.

*HE, C., J. RIGGS, C. S. HE, J. F. RIGGS, AND Y. T. KANG. 1993. Integration of geographic information systems and a computer model to evaluate impacts of agricultural runoff on water quality. *Water Resources Bulletin* 29(6):891–900. This study integrates an Agricultural Non-Point Source Pollution Model (AGNAPS), the Geographic Resource Analysis Support System (GRASS), and GRASS WATERWORKS to evaluate the impact of agricultural runoff on water quality in the Cass River, a subwatershed of Saginaw Bay.

*HUNSAKER, C. T., AND D. A. LEVINE. 1995. Hierarchical approach to the study of water quality in rivers. *Bioscience* 45(3):193–203. Explores how land-use data can be best used for modeling water quality. The authors suggest using a two-step process to identify areas in which management actions will be most effective.

KAUFMAN, L. 1993. Catastrophic change in species-rich freshwater ecosystems. *Bioscience* 42:846–858.

LATHROP, R.C., S. R. CARPENTER, C. A. STOW, P. A. SORANNO, AND J. C. PANUSKA. 1998. Phosphorus loading reductions needed to control blue-green algal blooms in Lake Mendota. *Canadian Journal of Fisheries and Aquatic Sciences* 55:1169–1178.

*LIKENS, G. E., AND R. H. BORMANN. 1995. *Biogeochemistry of a Forested Ecosystem.* Springer-Verlag, New York. Provides an overview of the Hubbard Brook Ecosystem Study begun in 1963 that has been foundational to the study of ecosystem processes at the scale of large watersheds and has been a pioneer in large-scale ecosystem experiments.

*MCINTOSH, R. P. 1985. Ecosystem ecology, systems ecology, and big biology. In R. P. McIntosh, ed., *The Background of Ecology: Concepts and Theory.* Cambridge University Press, Cambridge, UK, pp. 193–241. An overview of historical thought of ecosystem ecology, systems ecology, and analysis and the International Biological Program.

NEWMAN, E. I. 1997. Phosphorus balance of contrasting farm systems, past and present. Can food production be sustainable? *Journal of Applied Ecology* 34:1334–1347.

NOWAK, P., S. SHEPARD, AND C. WEILAND. 1996. Utilizing a needs assessment in water quality program implementation for the Lake Mendota watershed. The Farm Practices Inventory (FPI) Report #2. Available from Environmental Resources Center, University of Wisconsin.

OMERNIK, J. M., A. R. ABERNATHY, AND L. M. MALE. 1991. Stream nutrient levels and proximity of agricultural and forest land to streams: Some relationships. *Journal of Soil and Water Conservation* 36:227–231.

OSBORNE, L. L., AND D. A. KOVACIC. 1993. Riparian vegetated buffer strips in water quality restoration and stream management. *Freshwater Biology* 29:243–258.

OSBORNE, L. L., AND M. J. WILEY. 1988. Empirical relationships between land use/landcover and stream water quality in an agricultural watershed. *Journal of Environmental Management* 26:9–27.

PETERJOHN, W. T., AND D. L. CORRELL. 1984. Nutrient dynamics in an agricultural watershed: Observations on the role of a riparian forest. *Ecology* 65:1466–1475.

REUTER, J. E., T. DJOHAN, AND C. R. GOLDMAN. 1992. The use of wetlands for nutrient removal from surface runoff in a cold climate region of California—Results from a newly constructed wetland at Lake Tahoe. *Journal of Environmental Management* 36:35–53.

*RUNNING, S. W., R. R. NEMANI, D. L. PETERSON, L. E. BAND, D. F. POTTS, L. L. PIERCE, AND M. A. SPANNE. 1989. Mapping regional forest evapotranspiration and photosynthesis by coupling satellite data with ecosystem simulation. *Ecology* 70: 1090–1101. Describes landscape-level simulation model of annual evapotranspiration and net photosynthesis in a mountainous region.

SCHINDLER, D. W. 1977. Evolution of phosphorus limitation in lakes. *Science* 195: 260–262.

*SCHLESINGER, W. H., J. A. RAIKES, A. E. HARTLEY, AND A. F. CROSS. 1996. On spatial pattern of soil nutrients in desert ecosystems. *Ecology* 77(2):364–374. Examines the spatial distribution and autocorrelation of soil nutrients (including N, P, and S) in the Chihuahuan desert of the southwestern United States as a function of invasion of semiarid grasslands by shrubs.

*SKLAR, F. H., AND R. COSTANZA. 1991. The development of dynamic spatial models for landscape ecology: A review and prognosis. In M. G. Turner and R. H. Gardner, eds. *Quantitative Methods in Landscape Ecology*. Springer-Verlag, New York, pp. 239–288. An overview of spatial modeling as approached by both the social and natural sciences.

U. S. ENVIRONMENTAL PROTECTION AGENCY. 1990. The Quality of Our Nation's Water. EPA 440/4–90–005. Washington, D.C.

U. S. ENVIRONMENTAL PROTECTION AGENCY, 1997. INDEX OF WATERSHED INDICATORS. EPA-841-R-97-010. WASHINGTON, D.C.

*VITOUSEK, P. M., J. D. ABER, R. H. HOWARTH, G. E. LIKENS, P. A. MATSON, D. W. SCHINDLER, W. H. SCHLESINGER, and D. G. TILMAN. 1997. Human alterations of the global nitrogen cycle: Source and consequences. *Ecological Applications* 7(3): 737–750. General synopsis of global sources and sinks of nitrogen and how human processes have altered these fluxes.

*YOUNG, R. A., C. A. ONSTAD, D. D. BOSCH, AND W. P. ANDERSON. 1989. AGNPS: A non-point source pollution model for evaluating agricultural watersheds. *Journal of Soil and Water Conservation* 44(2):168–173. A computer model to analyze nonpoint source pollution and to prioritize potential water quality problems in rural areas.

APPLIED LANDSCAPE ECOLOGY
INTEGRATING ACROSS THE LEVELS

Landscape ecology is an exciting discipline in part because of the novel concepts and tools it brings to solving applied ecological problems. Landscape ecology makes important contributions to applied disciplines such as conservation biology and ecosystem management. We hope that the applications of landscape ecology have been apparent throughout this book, particularly in Chapter 4, Introduction to Markov Models; Chapter 5, Simulating Changes in Landscape Pattern; Chapter 7, Understanding Landscape Metrics I; Chapter 13, Interpreting Landscape Patterns from Organism-Based Perspectives; Chapter 14, Landscape Context; Chapter 15, Landscape Connectivity and Metapopulation Dynamics; Chapter 16, Individual-Based Modeling: The Bachman's Sparrow; and Chapter 18, Modeling Ecosystem Processes. In this section we present two chapters that apply the principles of landscape ecology to the conservation of natural areas. Chapter 19, Reserve Design, allows for the manipulation of simulated landscapes using a Reserve Design model to determine the optimal configuration of reserves for the persistence of different species. This lab examines the relevance of the spatial location and connectivity of reserves for the persistence of species with differing dispersal capabilities and life-history characteristics in the face of a dynamic disturbance regime. Chapter 20, Prioritizing Acquisitions for Conservation, enables students to use simple computer algorithms to select areas for preservation based on the current characteristics of sites, such as the species diversity or area, and examines many of the trade-offs that must be considered when selecting areas for conservation.

CHAPTER 19

RESERVE DESIGN

STANLEY A. TEMPLE AND JOHN R. CARY

OBJECTIVES

As the remaining natural areas of the world continue to shrink, it is increasingly important to protect representative examples that have the capacity to retain their typical biological diversity in perpetuity (Meffe and Carroll, 1997). This challenge forces conservationists to make appropriate choices when they select and design these protected natural areas. This exercise has three objectives related to this challenge. They are

1. to demonstrate how the tools of landscape ecology can be used to address crucial issues in conservation biology;
2. to demonstrate the challenges in selecting and designing protected natural areas to preserve a region's biodiversity; and
3. to show how design criteria, such as size, shape, and proximity to other natural areas, influence the population persistence of a variety of different organisms.

You will use a computer model, NATURE RESERVES, to explore a variety of selection and design issues that are important for conservation biologists and land-use planners to consider when they create protected natural areas. The model allows you to manipulate several key landscape features (e.g., proximity and size of patches) and see how your decisions affect the persistence of a variety of hypothetical species with challenging life-history characteristics. You will also use the model to explore the conservation potential of your own regional reserve system. Although the simulation that forms the basis for this exercise is based on a hypothetical situation, the choices you will be required to make are similar to the decisions conservation biologists make when selecting and designing protected natural areas. This lab requires a PC and the files accompanying this exercise on the CD. In addition, maps indicating the local parks and reserves for your region are needed for Exercise 3.

INTRODUCTION

Principles of Reserve Design

Conservation biologists are typically forced to make compromises when they select and design natural areas, such as parks and reserves, that are expected to preserve a representative sample of a region's biodiversity. Often, the major challenge is to use selection and design criteria that maximize the chance of population persistence for the species of concern in a region, while minimizing the total area that must be conserved or preserved. A variety of recommendations for preserve selection and design have been developed (e.g., Diamond, 1975; Noss and Cooperrider, 1994); some of the widely used guidelines that we will explore in this exercise are listed here:

- Preserves should be selected to capture the full range of biological diversity in an area.
- Preserves should usually be designed to be as large as possible.
- Several redundant preserves are better than a single preserve.
- All else being equal, a single large preserve is usually better than several smaller preserves totaling the same cumulative area as the large preserve.
- Preserves that are connected or close enough to allow organisms to move between them are better than preserves that are farther apart.
- Preserves that are compact in shape and minimize contact with dissimilar surrounding areas are better than preserves that are elongated or indented in shape and increase boundary (or edge) effects.

Although these and other guidelines exist (Shafer, 1991; Noss and Cooperrider, 1994), it frequently proves challenging to apply theses guidelines in real-world situations, as they may collide with competing land-use interests. Nonetheless, conservation biologists try to apply guidelines as closely as possible because it will help ensure that the benefits for biodiversity are maximized while societal costs are minimized. Many of the principles of landscape ecology can play a central role in the process of applying general guidelines to specific landscapes.

The previous guidelines have been developed to help contend with a variety of issues and characteristics of different species, as outlined in the following section. A major goal is to support enough individuals of a species to form a viable population. A **viable population** is a population that has a high probability of persisting over a long time period (Shaffer, 1981; Soule, 1987). In this lab you will examine several species characteristics and habitat affinities that are important to consider in the application of the reserve design guidelines.

Characteristic Habitat. Most organisms have a characteristic habitat in which they achieve high levels of survival and reproduction. Outside of the ecosystems that provide their optimal habitat, organisms do poorly and populations may fail to maintain themselves over time. **Habitat specialists** use a narrow range of ecosystem types, whereas **habitat generalists** successfully occupy many different ecosystem types. These habitat affinities are important in determining where a species will be found on a landscape and, hence, where preserves should be located.

Area Sensitivity. Some species are area sensitive and can only maintain themselves in large patches of habitat. Such species require large patches of habitat because often they live at such low densities that only large areas can support enough individuals to form a viable population. Such area-sensitive species include large organisms that are rare because of their size, top predators that are rare because of their position in the food chain, specialists that are rare because of their reliance on scarce resources, and migratory species that are area sensitive because they roam over large areas during their life cycle. The design challenge is often to accommodate the needs of the most space-demanding organisms in the community and use them as **umbrella species,** assuming that if their needs are met, the needs of most of the less space-demanding organisms will be met as well.

Isolation Sensitivity. Some species are isolation sensitive because they are narrowly restricted to a particular ecosystem and are easily marooned when a habitat patch becomes surrounded by a different ecosystem type (Hudson, 1991). Isolation-sensitive species may be small organisms with limited dispersal abilities, species that perish when they attempt to disperse across alien landscapes, and species with behavioral inhibitions to venturing into strange environments. Often an isolation threshold is set by the maximum dispersal distance of the organism across an alien landscape. The maximum dispersal distance can also vary depending on the characteristics of the habitat a species must cross. If preserves are separated by distances greater than this isolation threshold, the populations in each preserve will be "closed," and they must be large enough to maintain viability intrinsically within the preserve. In contrast, if preserves are separated by distances less than an organism's isolation threshold, the populations in adjacent preserves may actually function as one interconnected metapopulation (a network of semi-isolated subpopulations with some movements of individuals among them), and population viability may be achieved by the entire metapopulation rather than each individual subpopulation (Hanski and Gilpin, 1997). From the design perspective, preserves that are close enough to one another to allow dispersal movements of area-sensitive species may make very large preserves less necessary.

Edge Sensitivity. Individuals of some species may avoid edge habitats where two different ecosystems (e.g., forest and field) meet, or they may suffer reduced survival and reproduction there (Saunders et al., 1991). Edge-sensitive species may have problems coping with the physical and biotic conditions that exist at edges. The width of the edge habitat that becomes unacceptable varies considerably for different species. It is important for edge-sensitive species to have enough **core habitat,** far enough from edges of a preserve, to achieve population viability. Preserves that have a high proportion of their area near edges may actually have far less habitat for an edge-sensitive species than the preserve's total area suggests; some may be entirely edge habitat. In general, preserves that are designed to be as compact as possible and to have few indentations in their borders (i.e., a low edge-to-area ratio) tend to maximize the proportion of their area that is core habitat.

Disturbance Sensitivity. Disturbance-sensitive species are those that respond negatively to, or even suffer catastrophic losses as a result of, natural (e.g., fire) and anthropogenic (e.g., logging) disturbances occurring in their habitat. For these species it is important to manage the spatial aspects of disturbance so that a viable segment of the population survives on the landscape. This can be accomplished by designing preserves in close proximity so that if a population of a sensitive species in one area is extirpated by disturbance, immigrants from the nearby preserve can easily recolonize the disturbed area. This phenomenon is known as the **rescue effect,** and it can play an important role in the long-term maintenance of a metapopulation occupying a network of protected natural areas.

Satisfying the ecological requirements of a variety of different species within one particular system of reserves demands that conservation biologists consider the size of the protected areas, the isolation among habitat patches, the amount of edge relative to core habitat, and the impact of disturbance regimes. Keep in mind that we will not be able to explore many other important ecological and social issues in this simplified exercise (Noss and Cooperrider, 1994).

The NATURE RESERVES Model

The NATURE RESERVES model allows the user to create a system of protected natural areas and explore the consequences of different selection and design decisions for the population viability of several species on a hypothetical landscape. In the model, we assume that areas outside of protected areas will be developed for other uses, incompatible with the persistence of any of the species (although in reality, developed areas are not always biological deserts, and some species may thrive there). After the user designs a system of reserves, the program assesses the design and displays the consequences for each species.

Your task is to use our computer model NATURE RESERVES to explore various ways of preserving as many of the ten species as possible in a hypothetical landscape that will be completely developed, except for the areas you protect. Your goal is to preserve all ten species in as small an area as possible.

You get to select the percentage of the landscape that will be protected. Obviously, the larger the total area protected, the easier it is to preserve more species. The decisions you must make about protecting natural areas are very similar to the ones real-world conservation biologists must make. Given the limitations on the amount of land that can be protected, you will want to consider the following issues: Where will the preserves be located on the landscape? How many preserves will be needed? What sizes will the preserves be? How close to one another will the preserves be? What shape will the preserves be? How will you cope with disturbance within each preserve?

The hypothetical landscape in which you will work has two natural ecosystems: (1) green fields, occupying 75% of the land area, and (2) red woods, covering 25%. The landscape encompasses four townships (i.e., 12 × 12 mile or 19.2 × 19.2 km). A 16-hectare grid system is superimposed on the landscape, indicating the minimum patch size of the landscape—thus, you will be restricted to protecting land in 16-hectare units. We chose this minimum size for parcels because 16 hectares (40 acres) is a very common parcel size on many North American landscapes dominated by private ownership.

The landscape is also occupied by ten species of animals, each of which has a characteristic distribution among the ecosystems, a sensitivity to habitat patch size, a sensitivity to interpatch distance, a sensitivity to edges, and a sensitivity to disturbance regimes. These ecological characteristics are summarized in more detail in Table 19.1. Each species is associated with either one or both of the ecosystems. None of the ten species can live permanently outside of its designated habitat.

MODEL INPUT

Protection Goal. This sets the proportion of the landscape that will be protected in reserves; the rest of the landscape becomes developed and is uninhabitable for the species in the model. This parameter is set via the menu bar.

Mode of Selection. This gives you the option of either random or custom selection of reserves. In random mode, the model randomly assigns patches for protection. In custom mode, you get to select the locations, sizes, and shapes of patches to be protected. The selection mode can be changed by clicking on a choice under the Select option on the menu bar. We include this option to highlight the fact that protected areas must be carefully selected and designed; randomly or haphazardly assembled networks rarely, if ever, succeed.

Characteristics of Reserves. The model allows you to select the location, size, and shape of reserves by dragging and clicking the mouse.

MODEL OUTPUT

Summary Statistics. These provide a useful summary of what has been protected (the percentage of landscape preserved, the number of reserves [or patches], and the habitat composition of reserves). This information is output automatically and displayed at the bottom of the screen.

TABLE 19.1
CHARACTERISTICS OF THE TEN SPECIES THAT LIVE IN THE
HYPOTHETICAL LANDSCAPE.

Species	Preferred Habitat	Area Sensitivity— Home Range Size (hectares per individual)	Isolation Sensitivity— Isolation Threshold Distance (km)	Edge Sensitivity— Distance from an Edge to Suitable Habitat (km)	Disturbance Sensitivity (Is the Species Extirpated by Disturbance?)
1	Green fields and red woods	10	5.64	0	No
2	Green fields and red woods	5	2	0.4	No
3	Green fields	10	2.8	0	Yes
4	Green fields	5	3.99	0	Yes
5	Green fields	2.5	1.4	0	No
6	Green fields	1	1.78	0.4	No
7	Red woods	10	5.64	0	Yes
8	Red woods	5	2	0	Yes
9	Red woods	2.5	2.82	0.4	No
10	Red woods	1	0.9	0	No

Assessment of Effectiveness of Reserves. The assessment displays in tabular format how many secure home ranges have been preserved for each species. This is provided by clicking the **Assess** option on the menu bar and is displayed in an on-screen table.

Secure Home Ranges. These are displayed on the map of the landscape where a species has secure home ranges. A secure home range is an area of habitat in which the species has the potential for long-term presence. By clicking on a species name, these areas are displayed on the landscape map.

The model makes a variety of assumptions. Population viability implies that a population living within a preserve is large enough to avoid the problems associated with small population size (Soule, 1987; Lande, 1993). For purposes of this exercise, we will assume that for each of the ten species a population of at least 200 individuals will be needed to achieve long-term viability. Determining the actual requirements for population viability for each species in a region is a complex undertaking beyond the scope of this exercise (Boyce, 1992). Species occur at a different population density (Table 19.1) and therefore reach the designated viable population size in different sized patches of habitat. Relatively area-sensitive species (e.g., species 1) have a large home range requirement and exist at low population densities (e.g., 10 ha per individual). They can only achieve a viable population size in relatively large areas of habitat (i.e., at least 20 km²). Relatively area-insensitive species (e.g., species 6) have smaller space requirements (e.g., 1 ha per individual) and can achieve a viable population in relatively smaller areas (i.e., at least 2 km²). For purposes of this exercise, we will assume that all ten species occur at their maximum density (i.e., carrying capacity) in every suitable patch.

Each hypothetical species has an isolation threshold beyond which it will not be able to disperse successfully from one patch of suitable habitat to the next (Table 19.1). For species with limited dispersal abilities (e.g., species 10), patches of habitat must be relatively near one another (no less than 0.9 km apart) if they are to accommodate an interconnected metapopulation. Subpopulations in separate patches of protected habitat will be isolated from one another, and they will have to achieve viable population size within the confines of each isolated patch of habitat.

Disturbance can be assumed to strike every patch of habitat, but in our model landscape, disturbance never affects more than one patch at a time. When disturbance occurs, the entire patch is disturbed. Each of the ten species (Table 19.1) is either unaffected by disturbance (e.g., species 2) or extirpated completely from areas that are disturbed (e.g., species 3). For those species that are sensitive to disturbance, you can assume that they recolonize disturbed areas immediately after the disturbance has passed. Recolonization comes from nearby patches of habitat that are not contiguous but are within the species' potential dispersal distance. We assume that when the habitat is recolonized, the population promptly reaches maximum density.

Instructions for Using the NATURE RESERVES Model

1. Copy the entire **Reserve Design** directory from the CD-ROM to your hard drive.

2. Remove the "Read-only" status of the files by selecting each file with a *single* click of the *right* (not left) mouse button. This will produce a Windows pull-down menu. Under **Properties**, then **Attributes**, remove the "Read-only" status of every file in the directory.

3. Open the program by double-clicking on the file Reserve.exe.

4. Click on the **Readme** file on the Files menu bar and familiarize yourself with details about the program.

5. Examine the various selection and design features. These are accessed by using the pull-down menu labeled **Design** on the main menu bar.

6. Click on **Set Protection Goal** to set the proportion of the landscape that you will be allowed to preserve. We recommend that you start out with

TABLE 19.2
A DESCRIPTION OF THE INPUTS AND OUTPUTS OF THE PROGRAM
RESERVE DESIGN.

Input or Output	Parameter	Description
Input (via menu bar)	Protection goal	Sets the proportion of the landscape that will be protected in reserves; the rest of the landscape becomes developed and is uninhabitable for the species in the model.
Input (via clicking on a choice under the Select option on the menu bar)	Mode of selection	Gives the option of either random or custom selection of reserves. In random mode, the model randomly assigns patches for protection. In custom mode, the user selects the locations, sizes, and shapes of patches to be protected.
Input (via clicking and dragging mouse)	Characteristics of reserves	Allows the user to select the location, size, and shape of reserves
Output (automatically displayed at bottom of screen)	Summary statistics	Provide a useful summary of what has been protected (percentage of landscape preserved, number of reserves, habitat composition of reserves).
Output (via clicking on Assess option on menu bar; displayed in an on-screen table)	Assessment of effectiveness of reserves	Displays in tabular format how many secure home ranges have been preserved for each
Output (via clicking on a species name; displayed on the landscape map)	Secure home ranges	Displays on the map of the landscape where a species has secure home ranges.

a relatively large protection goal of 30%, and after you are familiar with the model, try progressively smaller protection goals until you can preserve all species in the minimum area.

7. Try some of the automatic options that allow the computer to randomly select preserves of a predetermined size. Click on **select random 160's,** which results in a random collection of 160-acre patches being preserved, up to the protection goal you specified. The screen displays the results on the map, and across the bottom of the screen there will be a series of useful statistics about the system of reserves you created (percentage of the landscape protected, number of reserves, habitat composition of reserves).

8. Once you have viewed the results of this random selection and design feature, you can see how well the ten species fared by clicking **Assess** on the main menu bar. The program will then analyze your design, and on the right-hand side of the screen it displays a species-by-species report of which species have secure populations of at least 200.

9. You can see where in your network of preserves each species is found by clicking on the species name on this list; the areas occupied will be highlighted in light blue on the map.

EXERCISES

EXERCISE 1
Random Reserve Design

Use RESERVE DESIGN to create several different random 64-hectare (160-acre) designs, and each time record which species have not been adequately protected. After several runs, you should be able to detect a pattern. (Use the default protection goal of 0.30 or 30%.)

Question 1. Are certain species never adequately protected by randomly located, 64-hectare (160-acre) preserves (i.e., you have failed to accommodate a viable population of at least 200)?

Question 2. Refer back to Table 19.1 and determine which characteristics were associated with the species not accommodated in a network of randomly selected 64-hectare (160-acre) reserves. Were they area sensitive, isolation sensitive, edge sensitive, or disturbance sensitive?

Repeat this in Exercise 1 process using the other random selection options that protect larger patches (either 256, 1024, or 4096 hectares). Each time, after you have made several runs of the model and noted which species were not accommodated, figure out why.

EXERCISE 2
Custom Reserve Design
---

1. Start by clicking **select none** from among the Design options. This option will result in the map showing no habitat protected, but you will still be able to see the distribution of the two habitats.

2. Create your own design by clicking the **mouse selects** option under the Design pull-down menu.

3. Use your mouse to create protected areas by clicking and dragging across the areas you wish to protect. Within the limits imposed by the selection goal, you can select as many protected areas as you wish

anywhere within the landscape, and they can be of any size or rectangular shape.

4. If you want to refine your design by trimming areas you previously selected, you can deselect areas by using the **mouse deselects** feature and clicking and dragging across the areas to be deleted.

5. Check the status report on the bottom of the screen for a summary of what you've protected.

6. When you are finished selecting and designing, click the **Assess** feature on the main menu bar to see how the ten species fared.

7. Repeat the process using a new design until you have been able to accommodate all ten species in 30% of the landscape.

8. Set the protection goal at 20% and repeat the process again. This time it will be more difficult to accommodate all of the species.

9. After you've succeeded with 20% of the landscape protected, reduce the percentage protected in 2% increments until you discover the absolute minimum area that can accommodate each of the ten species with a viable population size of at least 200.

Now that you have gone through the process of selecting and designing protected natural areas, you should be able to revisit the guidelines presented in the introduction and verify for yourself why they are so important.

Question 3. What design features are necessary to accommodate the most area-sensitive species (species 1, 3, and 7)?

Question 4. What design features are necessary to accommodate the most isolation-sensitive species (species 5, 6, and 10)?

Question 5. What design features are necessary to accommodate the edge-sensitive species (species 2, 6, and 9)?

Question 6. What design features are necessary to accommodate the disturbance-sensitive species (species 3, 4, 7, and 8)?

Question 7. What is the absolute minimum area required to support viable populations of all ten species?

EXERCISE 3
Your Regional Reserve System

Lack of attention to the principles of preserve selection and design typically results in a system of protected natural areas that cannot fulfill the goal of protecting biodiversity. Consider the protected natural areas in your region. Most road maps and atlases will show the main conservation areas, such as national or state parks, national or state forests, and wildlife refuges. Pick an actual area of the same size (12 × 12 miles) as the virtual landscape you have been working in during this exercise and note the locations, sizes, and shapes of protected areas. Answer Questions 8 and 9 on the next page.

Recreate the protected areas in your region in the model landscape and use the program to assess those protected natural areas to see how well they preserve the ten hypothetical species. Try to approximate the relative locations, sizes, and shapes of reserves as closely as you can. You will have to accept the distributions of the model landscape's two habitat types for now, even though they will not be relevant to your region.

Question 8. Do you see any clear evidence that the principles of selection and design explored in this chapter have been used in your region?

Question 9. If the answer to Question 8 is no, what criteria, if any, seem to have been used by the conservationists who selected and designed the protected natural areas in your region?

Question 10. What proportion of your regional landscape is actually protected?

Question 11. If the land use in the matrix between protected natural areas in your region eventually deteriorates into nonhabitat for many species, will the current network of protected areas support viable populations of area-sensitive, isolation-sensitive, edge-sensitive, and disturbance-sensitive species?

Question 12. Do some protected natural areas in your region lack some of these species because of inadequate selection and design?

Question 13. Think about top predators that are frequently area sensitive. How many protected areas in your region still have the largest native predators, such as wolves or bears, represented in the biotic community?

Question 14. What general changes might you recommend for the system of protected natural areas in your region?

CONCLUSIONS

Throughout this exercise you have made selection and design decisions within a small landscape while affecting only a few species. Even within this simple system, making the optimal decision for several species simultaneously can be challenging. It is even more challenging for real-world conservation planners to accommodate really challenging species, such as large top predators, in landscapes where competing interests can be powerful advocates for the development of natural areas. Time is short in many regions for conservation planners to select and design protected natural areas. Within a matter of a few decades, the options for protecting natural areas in many regions will be very limited as land is dedicated to other uses. For the future of biological diversity, it is crucial for planners to employ the principles introduced in this chapter in their decision-making processes. Failure to do so will have irreversible consequences for the biodiversity of many of the world's regions.

BIBLIOGRAPHY

Note. An asterisk preceding the entry indicates that it is a suggested reading.

BOYCE, M. S. 1992. Population viability analysis. *Annual Review of Ecology and Systematics* 23:481–506.

DIAMOND, J. 1975. The island dilemma: Lessons of modern biogeographic studies for the design of natural preserves. *Science* 193:1027–1029.

HANSKI, I. A., AND M. E. GILPIN, EDS. 1997. *Metapopulation Biology: Ecology, Genetics and Evolution.* Academic Press, San Diego, California.

HUDSON, W. E., ED. 1991. *Landscape Linkages and Biodiversity.* Island Press, Washington, D.C.

*HUNTER, M. L. 1996. *Fundamentals of Conservation Biology.* Blackwell Science, Cambridge, Massachusetts. A good general overview of the science of conservation biology. Chapters 11 and 14 are especially relevant.

LANDE, R. 1993. Risks of population extinction from demographic and environmental stochasticity. *American Naturalist* 142:911–927.

*MEFFE, G. K., AND C. R. CARROLL. 1997. *Principles of Conservation Biology,* 2nd ed. Sinauer Associates, Sunderland, Massachusetts. A more advanced and comprehensive review of conservation biology. Chapters 9 and 10 are particularly relevant.

*NOSS, R. F., AND A. Y. COOPERRIDER. 1994. SAVING NATURE'S LEGACY: PROTECTING AND RESTORING BIODIVERSITY. Island Press, Washington, D.C. A general review of the challenges conservation biologists face when trying to protect biodiversity.

SAUNDERS, D. A., R. J. HOBBS, AND C. R. MARGULES. 1991. Biological consequences of habitat fragmentation: A review. *Conservation Biology* 5:18–32.

SHAFER, C. L. 1991. *Nature Reserves: Island Theory and Conservation Practice.* Smithsonian Institution, Washington, D.C.

SHAFFER, M. L. 1981. Minimum population sizes for conservation. *Bioscience* 45:80–88.

SOULE, M. E., ED. 1987. *Viable Populations for Conservation.* Cambridge University Press, New York.

*STEIN, B. A., L. S. KUTNER, AND J. S. ADAMS, EDS. 2000. *Our Precious Heritage: The Status of Biodiversity in the United States.* Oxford University Press, New York. A thorough analysis of the current status of biodiversity in the United States and the challenges of creating enough protected natural area.

CHAPTER 20

PRIORITIZING RESERVES FOR ACQUISITION

DEAN L. URBAN

OBJECTIVES

Selecting sites for inclusion in a system of nature reserves typically entails preserving the most species in the least area or in smallest number of reserves. The reserve selection process exemplifies many of the challenges inherent to conservation in general, such as incomplete information, limited resources, and the need to balance competing goals. The goals of this exercise are to

1. introduce students to the computational logic of the site selection problem;
2. address issues of optimal as compared to suboptimal solutions; and
3. examine the trade-off between flexibility and adaptability versus quantitative rigor in the site selection process.

In Exercises 1–3 you will assemble hypothetical nature reserve systems according to three different selection algorithms, emphasizing either the total number of species, rare species, or landscape connectivity. In Exercise 4 you will extend these algorithms into a more flexible (and subjective) approach using the program PORTFOLIO, which allows for the consideration of multiple selection criteria. This lab requires a PC, the data files on the CD-ROM under the directory for this chapter, and printed copies of the tables from the CD.

INTRODUCTION

The express goal of a system of nature reserves, according to the Nature Conservancy (1996), is to preserve "multiple viable populations of all representative species" in a region. The reserve design process can be considered at multiple scales; indeed, the Nature Conservancy specifically addresses sites, landscapes, and networks of reserves in its planning efforts (Poiani et al., 2000). Reserve system planning unfolds in three stages. In the first stage, a large number of sites is screened for potential candidates as reserves. In the second stage, these candidate sites are examined for their promise as part of a functional system of reserves, or *portfolio*. In the third stage, the individual sites are actually established, managed, and monitored—a process that continues for the life of the reserve system. This lab exercise addresses the second stage of this process, the assembly of a reserve portfolio at the landscape scale.

The design of nature reserve systems uses multiple criteria in prioritizing sites for acquisition. The main selection criteria for the Nature Conservancy include the following:

Ecological uniqueness, which emphasizes the species richness or rarity of the taxa at a site, but also considers endemism or other ecological factors of special concern at the site.

Viability, which considers the likelihood that the species would persist on the site if protected. Factors contributing to the viability of a site involve various aspects of habitat "quality" such as reserve size, connectedness among sites, edge effects, and so on.

Threats, which broadly encompasses any natural or anthropogenic agents that might reduce the site's long-term viability or value as a reserve. Development pressure or impacts from surrounding land uses often constitute primary threats to nature reserves.

Feasibility, which includes a range of economic, sociological, and administrative factors related to the likelihood that the site can be acquired and protected.

Exactly *how* these criteria are considered in site selection—which criteria to emphasize, how to combine multiple criteria—is a matter of some flexibility and subjective choice. However, the site selection process can be aided by the use of selection algorithms, which are merely a set of rules for prioritizing candidate sites based on some selection criteria.

There are two general categories of selection algorithms (Csuti et al., 1997). **Heuristic algorithms** are typically iterative, in which an initial site is selected and then a second site is selected to maximize the gain in some criteria relative to what is already represented in the first site. Subsequent sites are then added accordingly. The algorithm is heuristic in that the selection that is made during each iteration in turn informs the next decision: the process "learns" as it proceeds. **Optimality algorithms,** by contrast, are solved simultaneously, often using computationally intensive linear-programming methods. While each approach has its advantages and drawbacks (Pressey et al., 1996, 1997;

Csuti et al., 1997), this exercise focuses on heuristic approaches because they are simple and intuitive and allow us to explore a broader range of alternative decision scenarios.

Here, you will examine three heuristic algorithms modified from Csuti et al. (1997). A **greedy richness** algorithm attempts to account for the most species in the minimum number of sites. This algorithm incorporates aspects of *complementarity*; that is, new species will tend to be accrued by adding sites that are species rich, but different from the sites already in the portfolio. A **greedy rarity** algorithm selects for the rarest species first, adding increasingly common species after the rare ones are accounted for. In other words, this is a greedy algorithm applied to rare species. A **connectivity** algorithm attempts to provide for landscape-scale population resilience by maximizing the likelihood of dispersal among sites by minimizing distances among sites or by providing stepping-stone habitats or dispersal corridors between sites.

The focus of these exercises is on the first two selection criteria, ecological uniqueness (specifically species richness and rarity) and viability (although certainly the other aspects of site selection are important!). You will use the algorithms to assemble a system of reserves under three "pure" selection scenarios: **greedy richness,** in which all species are treated the same using the greedy algorithm; **greedy rarity,** in which species rarity scores are used instead of simple occurrence; and **connected area,** in which a maximally connected set of reserves is assembled. For the first three exercises, the specific task is to assemble a portfolio of reserves according to each of these algorithms which maximize different features in the reserve system. In Exercise 4 you will use the computer program PORTFOLIO to aid in more complicated decision making. The ultimate goal is to consider the different solutions in terms of the reserve design criteria used in site selection.

Study Site

The study area comprises an imaginary landscape, a grid of 50×50 cells, containing ten sites being considered for inclusion in a portfolio of reserves (Figure 20.1). Table 20.1 shows the area of each site and as well as its **core area,** here defined as the area excluding "edge" cells in the raster map. A total of ten species of special concern have been censused in these sites (Table 20.2). These species are hypothetical but assume a range of abundances that will serve this exercise nicely. The species range from moderately common (eight sites occupied) to quite rare; two species were encountered on only a single site each. In general, larger sites support more species, and sites with proportionally more core area (less edge) have more species than expected for their size. The sites overlap in species composition to varying extents, suggesting that various combinations of sites will retain different species assemblages. Because some algorithms for portfolio design also consider the distances between candidate sites, a matrix of pairwise edge-to-edge distances between sites is also provided (Table 20.3).

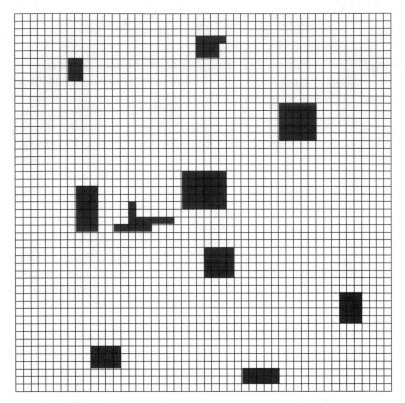

FIGURE 20.1
Map of ten potential sites under consideration for a portfolio of natural reserves
aimed at providing maximum diversity and maximum viability.

EXERCISES

EXERCISE 1
Greedy Richness Algorithm

The **greedy richness algorithm** begins by selecting the richest site as the initial reserve. The next reserve added to the portfolio is the site that adds the
most new species to the set of species represented in the initial site. In general, this new site will be a site maximally different in composition from
the first site (unless the new site is trivially poor in species). The third site
will be the site most different from both of the sites already in the set after the second step, and so on. Additional sites are added until all species
have been included on at least one site.

TABLE 20.1
DESCRIPTIONS OF CANDIDATE SITES.

Site	Area (ha)	Core (ha)	Rarity Score
1	6	0	
2	10	1	
3	25	9	
4	18	4	
5	13	0	
6	30	12	
7	16	4	
8	12	2	
9	10	0	
10	12	2	

While there are computer-based algorithms to do this, for a small number of species and potential sites the selection usually can proceed by inspection or "by-hand," as you will undertake next.

1. Prioritize the sites in Table 20.2 according to the greedy richness algorithm. Enter each site into Table 20.4 from the CD in the order it was selected, where step 1 is the initially selected reserve, proceeding through step 10 (the last reserve selected).

2. For each step, determine the following and enter into Table 20.4:
   (a) Cumulative area
   (b) Total number of species included in the reserves selected thus far (Cumulative Species Richness)
   (c) List of the new species included with each step

TABLE 20.2
CENSUS OF TEN SPECIES ON TEN POTENTIAL RESERVE SITES
(1 = PRESENT, 0 = ABSENT).

	Species									
Site	Sp 1	2	3	4	5	6	7	8	9	10
1	1	0	1	0	0	0	0	0	0	0
2	0	1	1	0	0	0	0	1	1	0
3	1	1	0	1	0	0	1	0	0	0
4	1	0	0	0	1	1	1	0	0	0
5	1	0	0	0	0	0	0	0	0	0
6	1	1	1	1	1	0	0	0	0	0
7	1	1	1	0	0	0	0	0	0	0
8	1	1	0	1	0	0	0	0	0	0
9	1	1	0	0	0	0	0	0	0	0
10	0	0	1	0	0	1	0	1	0	1

Tie-Breaker Decisions in the Algorithms

Cases might arise in which more than one site will meet the conditions of the algorithm (e.g., yield the same number of new species). In these cases, some sort of tie-breaker must be invoked. This presents an opportunity to create hybrid algorithms, using tie-breaker rules from other algorithms. For example, ties in the greedy richness algorithm might be decided on the basis of rarity or connectedness. Criteria such as threats or feasibility (cost) also might enter into the decision process at this point.

TABLE 20.3
PAIRWISE DISTANCES BETWEEN TEN POTENTIAL RESERVE SITES.

Site	1	2	3	4	5	6	7	8	9	10
1										
2	15.0									
3	26.3	10.4								
4	14.0	21.8	25.0							
5	19.3	21.4	17.6	2.0						
6	18.1	15.0	8.4	11.0	1.8					
7	27.6	25.0	15.6	14.3	5.4	5.0				
8	35.1	39.5	34.6	16.0	15.0	20.0	13.9			
9	43.8	41.2	30.0	26.6	26.6	21.2	12.2	16.0		
10	44.5	5.2	20.4	33.2	24.1	19.0	14.3	29.3	10.4	

EXERCISE 2
Greedy Rarity Algorithm

The greedy rarity algorithm proceeds in the same manner as the greedy richness algorithm, except that a rarity-based weight is used instead of simple species presence. One way to weight species uses a rarity score (which represents the inverse of their commonness) of $1/f$, where f is the frequency (or fraction) of sites where that species was recorded. Thus, a species that occurs on only one of the ten sites is assigned a weight of 10, a species that occurs on two sites is assigned a weight of 5, and a species that occurs on all sites is assigned a weight of 1. Using this approach, the sum of all the rarity scores for all the species at a site constitutes the richness tally for the site. This algorithm tends to capture rare species first, subsequently adding the more common species. The steps are essentially the same as for the greedy richness algorithm, but substituting the summed rarity scores for the richness tally of each site.

1. Compute the rarity score for each species on each site, and then determine the summed score for each site.
2. Rank the reserves based on the summed rarity scores and enter each site into Table 20.5 in the order in which they are selected.
3. At each step,
   (a) determine the cumulative area of all sites,
   (b) determine the cumulative number of species protected,
   (c) determine the cumulative summed rarity score for the reserve system, and
   (d) update the list of which species are included in the portfolio.

EXERCISE 3
Connectedness Algorithm

The connectedness algorithm focuses on viability rather than ecological uniqueness as the main criterion for site inclusion. This approach attempts to accrue a set of sites that are as funcionally connected as possible (Noss, 1991; Taylor et al., 1993) and consistent with metapopulation theory (Hanski and Gilpin, 1998). The selection begins with the largest site. The second site added is the one that adds the largest "connected area" to the initial site. For purposes of this exercise, the dispersal flux (or connectivity, f_{ij},) between sites i and j of area a can be indexed as

$$f_{ij} = a_i \cdot a_j \cdot p_{ij} \tag{1}$$

where the probability of dispersal, p_{ij}, between sites i and j is

$$p_{ij} = e^{(\theta \cdot d_{ij})} \tag{2}$$

where d_{ij} is the distance between sites i and site j. The parameter θ defines a negative-exponential dispersal-distance function. For this exercise, we will use $\theta = -0.1535$, which yields a function that tails to a dispersal probability of 0.01 at a distance of approxmately 30 map units.

The subsequent sites added are those that add the greatest connectivity. This algorithm thus "grows" a maximally connected reserve area, C.

$$C = \sum_{i=2}^{n} \sum_{j=1}^{i-1} a_i \cdot a_j \cdot p_{ij} \tag{3}$$

where the summation is over nonredundant pairs of sites in the set of as many as n sites. This solution is the parsimoniously connected "backbone" of the landscape (Urban and Keitt, 2001).

1. Use the connectedness algorithm to prioritize the ten reserves. For the first site, simply choose the largest site.

2. For the second site, choose the site j such that f_{ij} is maximized over all j not already in the reserve portfolio when site j is connected to *any* site i already in the portfolio.

3. For subsequent sites, continue to select the site that adds the greatest connectivity.

4. Enter the sites as selected into Table 20.6, and determine the following:
   (a) Cumulative total area
   (b) Cumulative total number of species included
   (c) Which additional species are included at each step
   (d) Distance-weighted connected area (C in the previous equation).

Note that as an alternative and more restrictive algorithm, this connectedness method could be constrained to focus on core area, for example, and then accrue the maximum connected core area. Some of the potential reserves have no core area and would be ineligible for inclusion by such an algorithm.

NOTE: The *greedy* rarity algorithm exercised here should not be confused with a simple *rarity* algorithm, in which the sites would be added in order of their overall rarity scores, that is, without reference to rare species already represented on previous sites. The simple rarity algorithm thus ignores complementarity and instead allows for *supplementarity*, in which redundant occurrences can increase overall viability. Returning to the Nature Conservancy's goal of "multiple viable occurrences of all elements [species or communities]" one might observe that the greedy richness algorithm focuses on the keyword *all*, the simple rarity algorithm keys on the word *multiple*, while the connectedness algorithm keys on *viable*.

Question 1. How do the algorithms compare in terms of their efficiency in accumulating species?
   (a) Compare the performance of the different algorithms by graphing cumulative species richness against the number of sites in the portfolio for each algorithm (with all three algorithms on the same graph).
   (b) Also, compare the performance of the algorithms by graphing cumulative species richness versus cumulative area.

Question 2. How would this assessment change if you compared the algorithms according to other criteria such as connected core area (instead of species richness)? *HINT:* Another graph is useful here.

Question 3. Considering total species richness, rarity, and viability as equally important criteria, is there an algorithm (or combination, or hybrid method) that will provide a "best" solution to this portfolio design? (To some extent, this will depend on how you elect to break ties in the selection algorithms.)

Indeed, it may have become clear that even in the extremely simple example developed in Exercises 1–3, the process of site selection can become woefully complex. Next, you will use the computer program PORTFOLIO to aid in more complicated decision making to assemble a reserve system iteratively.

Program PORTFOLIO

PORTFOLIO is a simple computer program for multicriteria decision support in assembling nature reserve systems and has been used in informal collaborations between the author and the Nature Conservancy in ecoregional planning efforts. The program has benefited from considerable input from the Nature Conservancy. In contrast to programs that attempt to maximally constrain or even optimize decisions, this program is designed to be maximally flexible; it merely provides information to a user, who may then use this information in a variety of ways (including the choice to ignore the information or make suboptimal decisions). The program computes four criteria used to compare candidate sites: habitat area, species richness, species rarity, and landscape connectivity. PORTFOLIO computes the *change* in each of these metrics that would be gained with the addition of each candidate site. These deltas are printed to the screen and the user is prompted to make a choice of the next site to include in the portfolio. The change in each of the metrics is then recomputed and the process iterates until the user terminates the session (typically, when all species have been accounted for in the portfolio).

The four selection criteria are pursued in PORTFOLIO by using several variations of the basic greedy algorithm:

> **Greedy richness,** with which you are already familiar from Exercise 1. Again, this approach emphasizes complementarity of sites.

> **Simple rarity,** in which sites are ranked according to total rarity of a site regardless of whether the species is already represented in the portfolio. This approach emphasizes supplementarity rather than complementarity.

> **Greedy rarity,** as defined in Exercise 2 (i.e., complementary rarity).

> **Connected area,** based on a somewhat more involved index of landscape connectivity based on graph theory (Harary, 1969; Urban and Keitt, 2001). PORTFOLIO computes a *minimum spanning tree* based on the dispersal probabilities indexed in equation (3), which is a measure of the relative dispersal distance spanning the reserve system. This index is consistent with the notions of "spreading-of-risk" or long-distance rescue effects as applied to metapopulations (den Boer, 1968; Levins, 1969). A highly connected reserve system would tend to be more resilient to chance events that decimated a local population (one or more sites) because such sites could be recolonized from sites farther away (Pulliam, 1988).

> **Connected core,** as defined earlier but considering only core habitat area.

To follow a "pure" strategy, always choose the site identified as providing the largest gain in the selected criteria. To follow a mixed strategy, choose sites based on any criteria at any stage of the selection process. Note that in many cases—including the contrived example included here—there may be ties at several steps of the process, in which cases other criteria may be invoked to make the selection. The program will alert the user in cases in which sites are tied on any selection criteria.

MODEL INPUT

The program requires three input data files. The **site file (site.txt)** lists the candidate reserve sites, with each site indexed by its area and core area (a proxy for habitat quality in the examples considered here). These are the data from Table 20.1. Optionally, this site file may also include ancillary variables indexing habitat heterogeneity, threat, and cost of each site. An example file with hypothetical ancillary data is included as **site2.txt.** This ancillary information is not used in the formal algorithms, but can be used in tie-breaking decisions.

The **distance file (distance.txt)** contains information on the spatial relationships among sites on the landscape, in terms of pairwise distances between sites. These distances might be simple Euclidean distances (edge-to-edge, in this example), or could be weighted according to the navigability of the intervening matrix (Gustafson and Gardner, 1996). Because a distance matrix is symmetric, the site distance file need only tally the lower triangle of the matrix, written in column form (i.e., the elements of Table 20.3 written in a column in order $d_{21}, d_{31}, d_{32}, d_{41}, \ldots$).

The **species file (species.txt)** includes alphanumeric codes for each species or conservation element. If rarity-based algorithms are of interest, user-defined rarity scores for each species can be included in this file.

The **census file (census.txt)** summarizes the occurrence of each species or conservation element on each of the sites (Table 20.2).

MODEL OUTPUT

The program writes two output files. A **summary file,** named by the user at run time, includes cumulative statistics for the growing portfolio. At each iteration (selection step), this file reports the site selected for addition or removal, the number of sites in the reserve system, and cumulative statistics (number of species, total rarity, connectivity, and so on) for the system.

A **log file** of the decision process echoes the interactive session in terms of all of the information provided to the user and what decisions were made at each step along the way. Thus, the decision process is self-documenting.

One final option the program provides is that the selection process can be run *backwards*; that is, sites can be removed in steps subsequent to their original inclusion. This flexibility can be especially important in connectedness algorithms (small stepping-stones that "bridge" larger constellations of habitat patches can be identified readily in a "backward-selection" approach: such a site will add little connectivity when added by itself but will cause a large decrease in connectivity if removed later, after it has bridged larger sites). More generally, the first selection of a set of complementary sites need not be the best one—and so PORTFOLIO offers the opportunity to "fine-tune" a set of sites. For example, you might discover that a site entered into the portfolio early in the assembly process can be removed in the final steps because its species have been provided in other sites added during the process.

The program is provided with documentation (in the portfolio\help folder in the directory for this chapter) and can be accessed with an Internet browser. The help documents describe the format of required data files, detail the run-time options and alternatives for the program, and include a sample session and example outfile files.

Instructions for PORTFOLIO

It is essential that you copy the entire Portfolio folder to your computer's hard drive. The program code, written in FORTRAN, is available for DOS-compatible PCs. The program runs by double-clicking on the **folio.exe** file or by typing "folio" at a DOS command prompt. The file **Readme.txt,** included in the Portfolio/doc directory, details how to install and run the program under Windows. The code was developed in a UNIX operating system (Sun Solaris) and has been ported to and tested under Windows NT 4.0. The program is a work in progress and is provided with no guarantees as to its robustness under changes of operating systems or compilers. For updated versions of the program, and for UNIX versions, contact the author or visit the Landscape Ecology Lab's website (http:// www.env.duke.edu/lel).

NOTE: When prompted by PORTFOLIO for the names of input data files, be sure to type the full name for the desired file, for example, "site.txt" or "census.txt."

EXERCISE 4
Site Selection as an Interactive, Multicriteria Decision Process

For this final exercise, you will use PORTFOLIO to assemble a reserve system. Limit your reserve portfolio to no more than three reserves and not more than 60 hectares total area. Use whatever criteria you want in your selection of reserves, but be prepared to explain and justify your rationale. NOTE: When prompted for the tail distance, enter "30."

Question 4. Justify which selection criteria were used and how ties were broken.

Question 5. Is there an obvious "optimal" solution for this simple system of reserves?

Question 6. How does the multicriteria decision process pursued with PORTFOLIO compare with the "pure" strategies done by hand?

Question 7. What other considerations would improve or confound the selection process as developed here for a hypothetical case? How might this process unfold for a real conservation application?

BIBLIOGRAPHY

Note. An asterisk preceding the entry indicates that it is a suggested reading.

*CSUTI, B., S. POLASKY, P. H. WILLIAMS, R. L. PRESSEY, J. D. CAMM, M. KERSHAW, A. R. KEISTER, B. DOWNS, R. HAMILTON, M. HUSO, AND K. SAHR. 1997. A comparison of reserve selection algorithms using data on terrestrial vertebrates in Oregon. *Biological Conservation* 80:83–97. A good overview of various selection algorithms.

DEN BOER, P. J. 1968. Spreading of risk and stabilization of animal numbers. *Acta Biotheoretica* 18:165–194.

GUSTAFSON, E. J., AND R. H. GARDNER. 1996. The effect of landscape heterogeneity on the probability of patch colonization. *Ecology* 77:94–107.

*HANSKI, I. A. 1998. Metapopulation dynamics. Nature 396:41–49. A concise overview of current ideas on metapopulations.

HANSKI, I. A., AND M. L. GILPIN. 1997. *Metapopulation Biology.* Academic Press, San Diego.

HARARY, F. 1969. *Graph Theory.* Addison-Wesley, Reading, Massachusetts.

*HARRISON, S. 1994. Metapopulations and conservation. In P. J. Edwards, N. R. Webb, and R. M. May, eds. *Large-Scale Ecology and Conservation Biology.* Blackwell, Oxford, pp. 111–128. A good overview of the theory that underlies much of current conservation biology. Blackwell, Oxford, England.

LEVINS, R. 1969. Some demographic and genetic consequences of environmental heterogeneity for biological control. *Bulletin of the Entomological Society of America* 15:237–240.

NOSS, R. F. 1991. Landscape connectivity: Different functions and different scales. In W. E. Hudson, ed. *Landscape Linkages and Biodiversity.* Island Press, Washington, D.C., pp. 27–39.

*POIANI, K. A., B. D. RICHTER, M. G. ANDESON, AND H. E. RICHTER. 2000. Biodiversity conservation at multiple scales: Functional sites, landscapes, and networks. *BioScience* 50:133–146. A state-of-the-art overview by the Nature Conservancy.

PRESSEY, R. L., H. P. POSSINGHAM, AND J. R. DAY. 1997. Effectiveness of alternative heuristic algorithms for idenitifying indicative minimum requirements for conservation reserves. *Biological Conservation* 80:207–219.

PRESSEY, R. L., H. P. POSSINGHAM, AND C. R. MARGULES. 1996. Optimality in reserve selection algorithms: When does it matter and how much? *Biological Conservation* 76:259–267.

PULLIAM, H. R. 1988. Sources, sinks, and population regulation. *American Naturalist* 132:652–661.

TAYLOR, P. D., L. FAHRIG, K. HENEIN, AND G. MERRIAM. 1993. Connectivity is a vital element of landscape structure. *Oikos* 3:571–573.

THE NATURE CONSERVANCY. 1996. *Conservation by Design: a Framework for Mission Success.* The Nature Conservancy, Washington, D.C.

URBAN, D. L., AND T. H. KEITT. 2001. Landscape connectivity: A graph-theoretic perspective. *Ecology* 82:1205–1218.

INDEX